SPIRULINA
WORLD FOOD

How this micro algae can transform your health and our planet

Updated and Revised

Robert Henrikson
www.spirulinasource.com
Ronore Enterprises, Inc., Richmond CA

Notice

This book is a reference guide to microalgae. It is intended to be solely educational and informational. It is not intended to sell any particular product. It is not intended as medical advice or as a guide to self-treatment nor is it intended to substitute for any treatment prescribed by a physician. People with medical questions should consult their doctor or health professional.

SPIRULINA
WORLD FOOD

Cover design collage by Robert Henrikson
Back cover photo of Robert Henrikson.

Library of Congress Catalog Card Number: 89-091683
ISBN 978-1-7376326-0-3

Printed in the United States of America.
Published by Ronore Enterprises, Inc.
PO Box 71024, Richmond CA 94807 USA
spirulinasource.com • roberthe@sonic.net

Earth Food Spirulina
First Printing, June 1989.
Second Printing, October 1989.
Third Printing, Revised Edition, April 1994.
Fourth Printing, Revised Edition, January 1997.
Fifth Printing, Revised Edition Online, November 1999.
Sixth Printing, Revised Edition Online, July 2009.

Spirulina World Food
Seventh Printing, Updated and Revised, September 2010.
Eighth Printing, Updated and Revised, July 2021.
spirulinasource.com • smartmicrofarms.com • algaecompetition.com

Acknowledgements

I wish to express my appreciation for the dedicated people who have pursued the dream of personal and planetary transformation by developing spirulina and other microalgae.

I would like to thank Larry Switzer, David Donnelley, Shigekuni Kawamura, Alan Jassby, Robert Bellows, Takemitsu Takahashi, Hidenori Shimamatsu, Terry Cohen, Bill Lackey, Yoshimichi Ota, Amha Belay, Juan Chavez, Ron Henson, Ernest Thomas, Paul Schofield, Mark Belock, Ripley Fox and many other contributors.

For their assistance in editing, I wish to thank my brothers John Henrikson, Richard Henrikson and Jack Wight. For recommendations in reformatting and typography for ebooks, David Punchard and Mark R. Edwards.

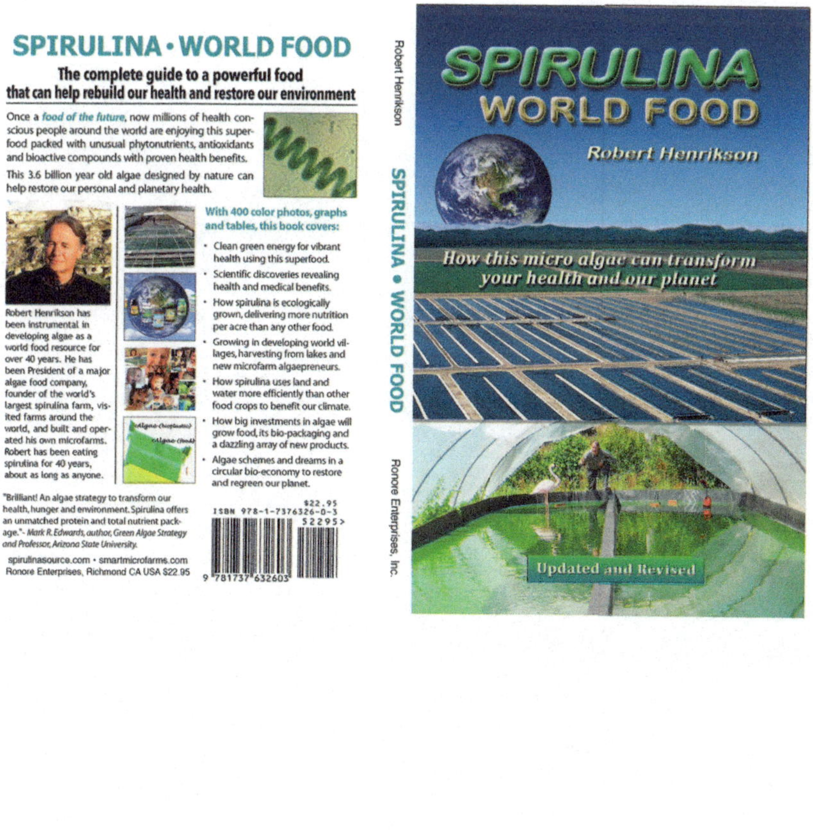

Table of Contents

Foreword by Larry Switzer

This book completes another step in the realization of a shared vision quest. Robert and I have been "Brothers of a Great Dream" for more than fifteen years. It is a dream no one owns but all can share – if they have the imagination and faith to rise to its promise, to know that it can and must come true. It is a Great Dream because we intuit that it is also the dream of the spirit that creates our destiny – the spirit of this living world. Our instincts tell us that this living world is not only the home of life, but is itself a gigantic, self-evolving organism some have called Gaia. She too has embarked upon a four billion year vision quest: to fill herself with conscious life.

We humans are an essential player in her Great Dream. In our awakening to her reality, we sense that she too is further awakening. Somehow Gaia herself is coming to fuller consciousness of her embryos of awareness called humans. Such is the true position of humanity – awakening to the full miracle of its most ancient and living parent, this living, conscious Earth.

If we fully accept the fact of her miracle, the fulfillment of her vision quest becomes ours. Our awakening invites us to care for our most ancient and sacred parent by co-creating upon her and with her. We are coming to realize that we can no more ravage and destroy her and survive than a foolish child can ravage and destroy its own mother. She asks us to honor her by honoring each other and the preciousness of life itself. And if we respond to her, we can be certain that she in turn will continue to be generous with us and with our far descendants.

We can awake to the startling good news that we can live and thrive within her miraculous being, that we can enjoy a sustainable sanity, peace, and abundance which until now has only been glimpsed by visionaries. This is how her vision quest is fulfilled in our awakening.

I honor this book because I know it was created in the service of her spirit. It reveals and explains one of the vital gifts the parent- Earth- offers its awakening child- humanity.

- Larry Switzer, May 1989.

Robert Henrikson & Larry Switzer.

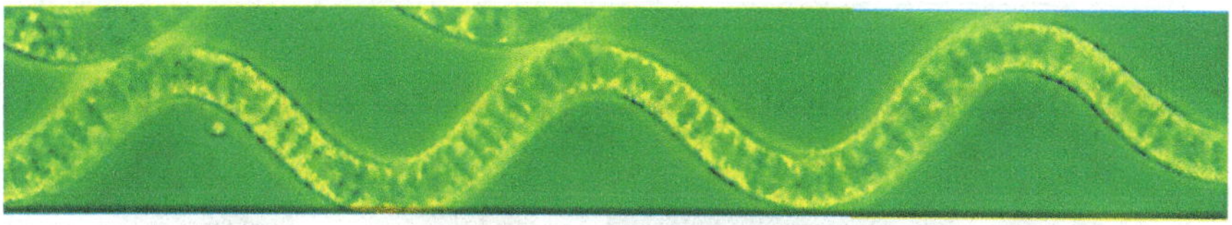

Spirulina, "little spiral"

Invocation

Spirulina speaks to the human species, on behalf of the first species - algae.

"For a billion years we filled Earth's atmosphere with enough oxygen so new lifeforms could evolve. We participated in the unfolding of diverse lifeforms on this planet for billions of years more. Paradise unfolded and we loved it.

Just moments ago, your species appeared. Over the past century, you humans have been shutting down life support systems of our planet. Your ability to upset the biosphere, deplete the ozone layer, increase global warming, deforest the land, expand deserts and pollute land, water and air has aroused our attention.

We do not depend on your gratitude for providing a beautiful planet, but a little cooperation is in order. Your plundering destroys the opportunity for life to unfold fully in all its forms. If you persist, your species will likely perish too. We algae will survive, and over the eons will again foster new life.

We do recognize your unique destiny on this planet. The future of planetary evolution rests with you. Heal yourself within, heal your relationships with your own species, and heal our planet. From this great challenge will emerge your highest creativity. Take this evolutionary jump and we want to participate with you.

We offer wisdom for your personal and planetary health. Embrace, befriend and learn from us. Rediscover the ancient wisdom of your ancestors." - ***Spirulina, circa 1989.***

This book is dedicated to Larry Switzer and Ripley Fox

Larry Switzer

Larry Switzer, a visionary bioneer and catalyst, founded Proteus Corporation in 1976 to develop spirulina blue-green algae as a world food resource. Proteus was funded by a group of California investors, committed to his spirulina vision.

Proteus Corporation with Dainippon Ink & Chemicals founded Earthrise Farms in 1981, which became the world's largest spirulina farm by the 1990s. In 1979 Larry launched the Earthrise® Spirulina product line which was expanded over the years and eventually distributed in 30 countries around the world.

Ripley Fox

Ripley Fox was passionately devoted to innovative solutions to combat malnutrition in the developing world and a pioneer of village scale spirulina production since 1969. Ripley developed integrated health and energy systems for village spirulina production.

Ripley authored *Spirulina Production & Potential* and numerous articles on spirulina. With his wife Denise Fox, Ripley inspired many humanitarian organizations working in the developing world, which led to the evolution of spirulina microfarms across France and Europe.

Introduction to a perfect food to restore our health

Our plunge into technology and our consumer lifestyle has upset the balance of Earth's biosphere. We transform Earth's resources into trash and pollution at a faster and faster pace. Emerging from ecosystem breakdown are super bacteria, viruses, chemical toxins, global warming and climate change that threaten our own health.

More people are becoming aware that diseases like cancer are directly related to environmental factors. A growing portion of Earth's population is seeking super health to protect against this pollution, boost their immune system, resist disease and retard the aging process.

All this arrives synchronistically at the beginning of a new millennium, when we sense a new age approaching: an age of networks, an age of community, or even an age of light. How can we make choices that restore our personal health, the health of our society and the entire planet?

Food choices for health restoration

1. *Does our food restore our personal health?*

Many foods are toxic to long term health, leading to cancer, heart and degenerative disease and immune breakdown.

2. *Does our food restore our human species?*

Much of the global food production and distribution system creates hunger in a world of abundance. Two-thirds of humanity live in poverty and scarcity.

3. *Does our food restore our planetary ecosystem?*

About three-quarters of all fertile land in the temperate and tropic zones is devoted to agriculture. Our world economy depletes fertile soil, wastes fresh water, pollutes the environment, turns rainforest into desert. The unsustainable way we produce food represents a great threat to our biosphere.

Planetary and personal health are under stress

Environmental stress accelerates the breakdown of Earth's ecosystems. These systems cannot adjust as quickly. Great bodies of water are dying from pollution. Forests are dying as if they were suffering from some immune system collapse.

Neither can human bodies quickly adjust to these rapid changes. Immune system diseases have appeared. Over prescription of antibiotics has accelerated the mutation of resistant bacteria. Antibiotics kill beneficial flora in the intestines, making room for infection by drug resistant bacteria.

The search for nutraceuticals to restore health

Researchers are scouring the globe for new foods and plants rich in disease preventing substances- antioxidants, nutraceuticals and designer foods with functional nutrients. Where do we start? With the original food designed by nature – spirulina!

Perfect food to restore our health and our planet

The first photosynthetic lifeform was designed by nature 3.6 billion years ago. Blue-green algae, cyanobacteria, is the evolutionary bridge between bacteria and green plants. It contained within it everything life needed to evolve. This immortal plant has renewed itself for billions of years. Spirulina has 3.6 billion years of evolutionary wisdom coded in its DNA.

Spirulina contains unique compounds like phycocyanin, polysaccharides, and sulfolipids that enhance the immune system. This superfood has the most remarkable concentration of functional nutrients ever known in any food.

On top of this, spirulina delivers more nutrition per acre than any other food on the planet. This has extraordinary implications for more efficient and less damaging food production.

Each day new research brings to light the wonders in microscopic algae. Research has shown phycocyanin and polysaccharide extracts of spirulina increase macrophage production, bone marrow reproduction, strengthen the immune system and disease resistance in fish, mice, chickens, cats and human cells. Spirulina contains sulfolipids, found to prevent viruses from either attaching to or penetrating into cells, thus preventing viral infection.

Algae is in its infancy as a food, medicine and biochemical resource. Spirulina was rediscovered about 55 years ago. Just 40 years ago, it burst into public awareness as a powerful new food with a promise to help feed the world's people.

Compared to five billion years of Earth history, or to millions of years of humanity, or to thousands of years of human food development, 40 years is an instant in time. Cultivation of grains and development of irrigation took thousands of years. Soybeans, a newcomer, took 50 years to emerge from obscurity. The last 40 years progress in algae technology is remarkable.

The third algae bloom has arrived

Over the past 20 years, billions of investment dollars from government, corporate and private sectors has poured into research and development for microalgae. This is the third great algae bloom.

In 1979, spirulina was introduced in the US health food market as a new natural food, gaining market share as an energizing, high-protein, all-natural, detoxifying food supplement. By the early 1980s algae had begun to take off as the *'food of the future'*. Even though production was less than 1,000 tons worldwide, still in its infancy, this represented the first algae bloom.

Over the next decade, scientific research documented the health benefits of spirulina and chlorella algae. Research on the immune system, detoxification and beneficial intestinal flora supported why people felt better. By the mid 1990s, the global market for algae food supplements was growing fast, backed by articles on health benefits. This was the second algae bloom. Big commercial algae farms got underway, especially in China and India.

After another decade, annual world microalgae output may have reached 10,000 tons including spirulina, chlorella, dunaliella and haematoccocus. Yet algae production costs have remained high, limiting the market to higher value nutraceuticals, food supplements and specialty feed supplements.

Billion dollar investments in algae represent the third great algae bloom. The goal of the investment is to develop disruptive technologies that lower costs to produce algae products that are cost competitive with conventional sources.

As production costs fall, new algae products are entering the marketplace, such as algae omega 3 oils, algae food and feed products and algae-based resins, biopolymers, and bioplastics, replacing fossil fuel chemical products. This coming decade will witness a dazzling array of new products from algae.

A complete guide to a powerful green food

(1) Role of algae in history and implications from its productivity.
(2) Nutritional attributes described and compared with other foods.
(3) How to use and benefit from this superfood.
(4) Scientific research reveals spirulina is a therapeutic food.
(5) The variety of products around the world.
(6) Technology for growing spirulina, large and small farms.
(7) Farms are blossoming in the developing world.
(8) Algae's impact on food, agriculture and climate.
(9) Algae projects to restore and regreen the face of our planet.

Just as our body is composed of billions of cells working together as a single being, billions of lifeforms on Earth are working together as one organism. By adopting ecological food choices for restoring our health, we restore our planet.

Algae transforms light to living matter more efficiently. Eat lighter, eat less. Raise your energy to embrace the pace of change in this age of transformation. Consume less, live lighter on the Earth. Participate in the unfolding story of earth rising.

The oldest organisms– the ones who gave us life– are back.

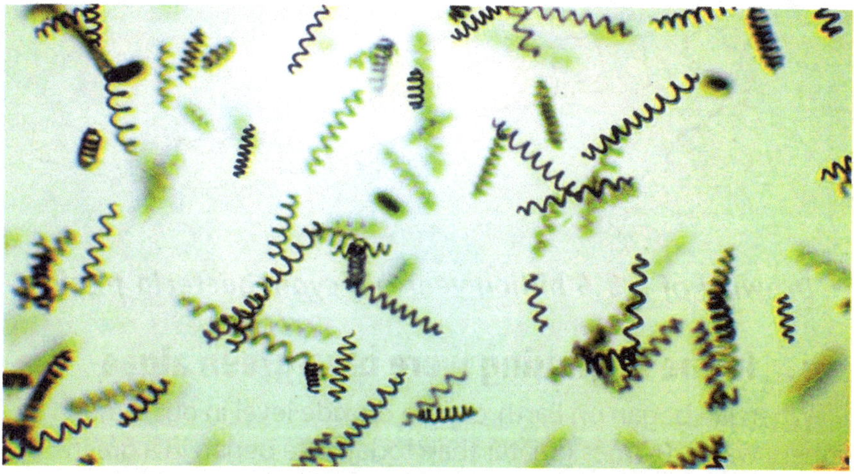

Chapter 1

Rediscovery of a 3.5 billion year old immortal lifeform

- **In the beginning ... were blue-green algae**
- **Algae through human history**
- **Human use – Legends, Kanembu and Aztecs**
- **Spirulina lakes and pink flamingos**
- **Thousands of algal species cover the Earth**
- **A new era of ecological agriculture**

Spirulina is the immortal descendent of the first photosynthetic lifeform. Beginning 3.5 billion years ago, blue-green algae created our oxygen atmosphere so other life could evolve. Since then, algae have helped regulate our planet's biosphere.

Algae are two-thirds of Earth's biomass. Thousands of species covering the Earth are now being identified for food, pharmaceuticals, biochemicals and fertilizers. Algae represent one of the solutions we need for food while restoring our planet.

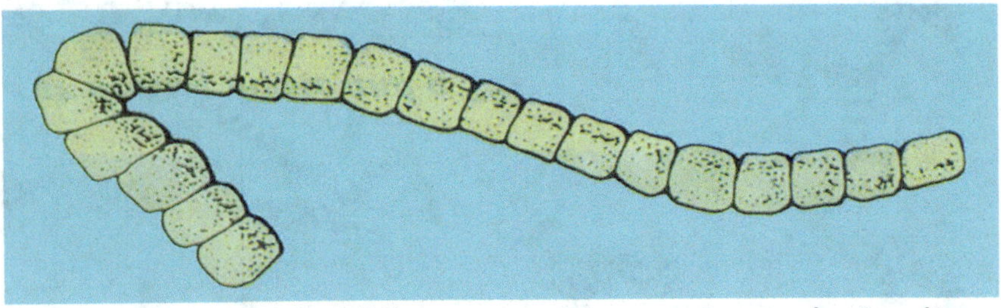

Drawing of a 3.6 billion year old cyanobacteria fossil.

In the beginning were blue-green algae

When life began on Earth, carbon dioxide level in our atmosphere was 100 times greater than today. Life began in a greenhouse atmosphere, and microalgae played the central role in transforming this inhospitable planet into the beauty and richness that makes up life today.

Scientists believe the Earth formed 4.5 billion years ago, and the first lifeforms appeared 3.6 billion years ago. The first bacteria, procaryotes, consumed chemical nutrients as food, but some adapted the energy of the sun to make their own food. The first photosynthesizing procaryotes, called cyanobacteria or blue-green algae, used light energy to break carbon dioxide and water molecules into carbon compounds, releasing oxygen. Fossils 3.6 billion years old show filaments of cells resembling spirulina.

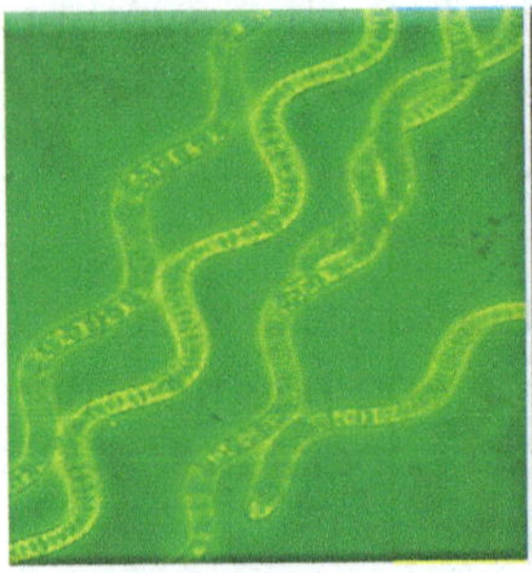

Filaments.

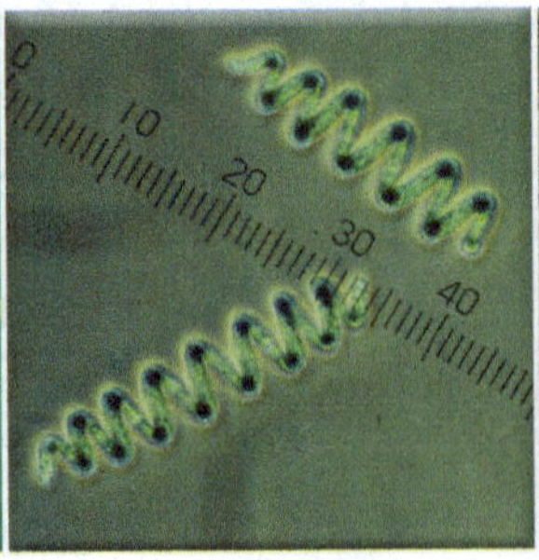

Spiral coils.

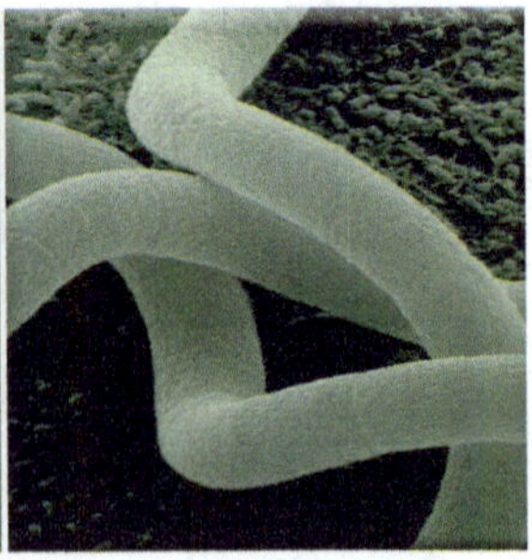

Enlarged.

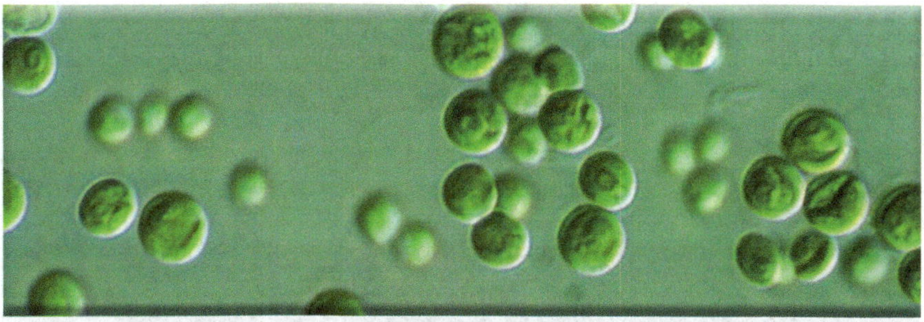

Chlorella, green algae with nucleus and cell walls.

Evolution of higher life forms

Iron and sulfur compounds in the oceans mopped up the free oxygen. Methanogen bacteria consumed decomposed algae and converted carbon to methane gas and carbon dioxide, compensating for the removal of carbon dioxide by photosynthesizing algae.

Over a billion years passed. When oxygen absorbing compounds in the oceans were used up, atmospheric concentration of oxygen increased rapidly. Methanogen bacteria retreated into the only environments devoid of oxygen – beneath the sea floor, in marshes, and in the guts of other organisms.

About 2.3 billion years ago oxygen may have reached a 1% level, and methane, a greenhouse gas, disappeared from the atmosphere, cooling the planet.

Cells with nuclei appeared. This more powerful lifeform was supported by higher oxygen levels. These eukaryotes, microscopic green algae, formed from communities of individual bacteria living within an outer membrane of one of them. The nucleus contained organelles such as chloroplasts. Because each organelle carried different genetic codes, the loss of information of one of them could mean the death of the cell. To overcome this possibility of death, sex evolved as a way to transfer information between cells.

Contributions of microalgae (David Punchard).

About 600 million years ago, Earth entered the present phase with the evolution of large plants and animals. The power requirements of larger organisms needed higher oxygen concentration which increased and stayed steady at 21%. For hundreds of millions of years, Earth's biosystem has kept the oxygen level balanced between 15%, where higher life forms cannot survive, and 25%, where forests would spontaneously combust in a global fire.

Procaryotes, cyanobacteria, or blue-green algae, still cover the land and water, part of the living mechanism for regulating the planet's biosphere. Our rediscovery of this original lifeform represents our need to return to the origins of life to heal our planet. Realizing that algae took billions of years to build and maintain the atmosphere, it is remarkable that humanity has raised the carbon dioxide concentration over 25% in merely one hundred years.

How important is the contribution this original lifeform? Brian Swimme, in *The Universe is a Green Dragon* writes:

> "We should take the procaryote as the mascot of the emerging era of the Earth. What better organism to symbolize the vast mystery of the Earth's embryogenesis.. Let's just hope we can emulate some of the achievements of the procaryotes ..To begin with, it would be wonderful if we could contribute something as essential to Earth's life as oxygen."[1]

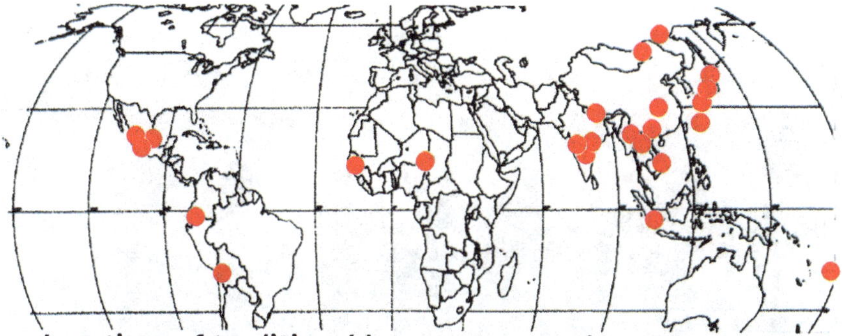

Locations of traditional human consumption of freshwater microalgae, excluding ocean seaweeds (Alan Jassby).

Algae in historical legends

The Bible describes when the Israelites were starving in the wilderness, God provided 'manna'– a flake-like thing, lying on the ground. They gathered the manna and baked it into bread. Some believe manna was a lichen – a combination of fungus and blue-green algae forming a crust on the rocks and ground.[2]

A thousand years ago in Vietnam, a monk discovered rice was more productive when a water fern, azolla, was planted in the paddies. Grateful farmers built temples to him but kept it secret. A woman named Ba Heng rediscovered azolla 700 years later. Growing rice with azolla continued for centuries, raising yields and saving many people from starvation. This century scientists discovered blue-green algae living on ferns fix nitrogen as biofertilizer.[3]

Although freshwater algae has not been eaten nearly as much as larger marine seaweeds, historical literature revealed at least 25 cases where 9 types of wild freshwater algae were collected and eaten in 15 countries.[4] This non-seaweed algae has been used in soups, spreads and sauces as a source of vitamins and minerals.

When microalgae could be collected because it formed larger colonies of mats or globules, it played a culinary and therapeutic role. Eating algae may have been limited only by the difficulty of collecting these tiny organisms.[5]

Kanembu Spirulina Ladies harvesting from Lake Boudou Andja.
Photos by Marzio Marzot from the FAO Report The Future is an Ancient Lake, 2004.

Dihé in Chad, Central Africa

In 1940, a journal published a report by a French phycologist Dangeard on *dihé,* eaten by Kanembu people near Lake Chad. *Dihé* is cakes of sun-dried blue-green algae collected from the shores of Lake Chad. Dangeard heard this same algae populated lakes in the Rift Valley of East Africa, and was the main food for flamingos there. His report went unnoticed.

In 1964, a French botanist on a Saharan expedition, Léonard, noticed blue-green algae in pools around Lake Chad. He came across blue-green cakes in markets of Fort Lamy (now Ndjemena). When locals said these cakes came from areas near Lake Chad, he recognized the connection between algal blooms and cakes sold in the market. He observed 70 percent of the food of the Kanembu was accompanied by a sauce made with these dried cakes.

Techniques of harvesting and drying have passed from mother to daughter for generations. Kanembu collect wet algae in clay pots, drain out water through bags of cloth and spread out algae in the sand to dry in the sun.

Ladies harvesting and drying spirulina dihé in a sand filter.

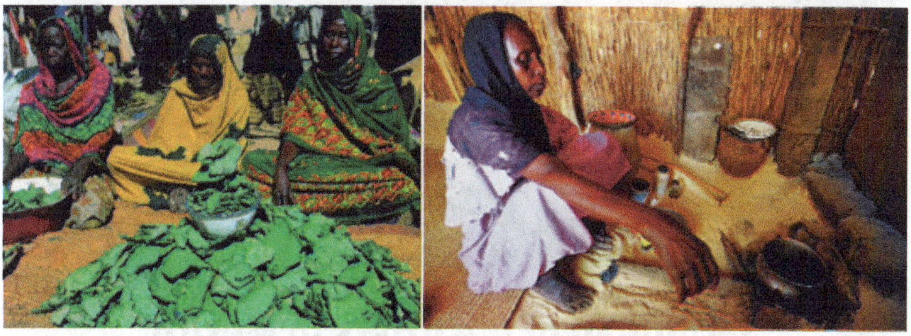

Selling dihé in the local market.

Photos by Marzio Marzot from the FAO Report The Future is an Ancient Lake, 2004.

When dry, women cut algae cakes into squares for sale in the local market. Dihé is crumbled and mixed with a sauce of tomatoes and peppers and poured over millet, beans, fish or meat, eaten in 70% of their meals. Pregnant women eat dihé cakes because they believe its dark color will screen their unborn baby from the eyes of sorcerers.[6] Spirulina is applied as a poultice for treating diseases.

The average consumption of dihé could be as high as 50 grams per person per week. More than 250 dry tons per year is produced, making the ladies of Chad certainly the lowest cost producer of spirulina in the world..

Aztecs harvesting blue-green algae from lakes in the Valley of Mexico.

Human Nature, March 1978. (by Peter T. Furst).

Techuitiatl of the Aztecs in Mexico

About the same time, a company director in Mexico, H. Durand-Chastel, read about spirulina and realized it was the algae clogging the soda ash extraction plant on Lake Texcoco.

An historical search revealed spirulina was harvested and used for human consumption 400 years earlier, at the time of the Spanish conquest. Chroniclers described fisherman with nets collecting *'techuitlatl'* from lagoons, and making bread or cheese from it. Legends say Aztec messengers took spirulina on their marathons.

Techuitiatl disappeared soon after the Spanish conquest. The lakes in the Valley of Mexico were drained for the new civilization. The only remnant today, Lake Texcoco, still has living algae culture.

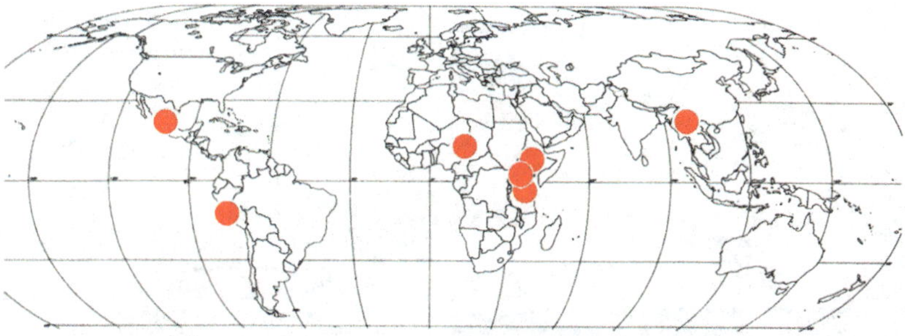

Lakes with natural spirulina blooms.

Spirulina lakes and pink flamingos

Besides Lake Texcoco, large spirulina lakes are in Central Africa and in East Africa along the Great Rift Valley. Under normal water conditions, spirulina may be one of many algal species. But the more alkaline and salty the water becomes, more inhospitable to other lifeforms, spirulina flourishes.

Lakes Bodou and Rombou in Chad have a monoculture of spirulina for centuries. It is a major species in Kenya's lakes Nakuru and Elementeita and Ethiopia's Aranguadi and Kilotes. The lesser flamingo evolved a filter in its beak to eat spirulina. Millions of flamingos feed entirely on algae when it is abundant.

Pink flamingos feeding on spirulina.

Collage from a Google search of images of microscopic algae.

Thousands of algae species cover the earth

There may be more than 25,000 species of algae, living everywhere. They range in size from a single cell to giant kelp over 150 feet long. Most algae live off sunlight through photosynthesis. Some live off organic matter like bacteria.

Microalgae can be seen under a microscope. Some break down sewage, improve soil structure and fertility or generate methane and fuels for energy. Others are grown for animal and aquaculture feeds, human foods, pharmaceuticals and biochemicals.

Larger algae, like seaweeds, are macroalgae. They already have an important economic role. About 70 species are used for human food, animal feed, fertilizers and biochemicals.

Microalgae in the ocean, called phytoplankton, are the base of the food chain and support all higher life. The rich upwelling of nutrients caused by the major currents meeting the continental shelf, or nutrients from river basins sustain phytoplankton growth.

Colors of algae (Mark Edwards).

There are blue-green microalgae like spirulina, green algae like chlorella and scenedesmus, red algae like dunaliella, and brown, purple, pink, yellow and black algae. They are everywhere – in water, in soils, on rocks, on plants. Blue-green algae are the most primitive, and contain no nucleus or chloroplast. Their cell walls evolved before cellulose, and are composed of mucopolysaccharides. Blue-green algae do not sexually reproduce; they divide.

Some blue-green algae fix atmospheric nitrogen into organic forms - essential for building proteins and amino acid complexes in plants and animals. Although nitrogen comprises 78% of the atmosphere, it is not usable by most plants and animals. For more productive crops, nitrogen must be added. Nitrogen comes from chemical fertilizers, microbial mineralization of organic matter, nitrogen-fixing bacteria in legumes, or blue-green algae.

Because it can fix nitrogen, blue-green algae is often the first life-form to colonize desolate land– in deserts, on coral reefs and polar regions, working with lichen to fix nitrogen to the rocks.

Spirulina, whose scientific name is *arthrospira platensis*, is an edible, non-nitrogen fixing blue-green algae. With a long history of safe human consumption and 50 years of safety testing, it meets all international food quality and safety standards. Specially designed farms grow spirulina under controlled conditions.

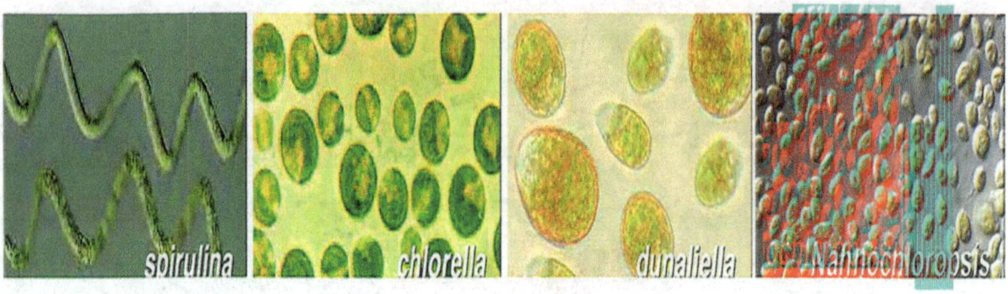

Chlorella for nutrition and protein

Chlorella was the first microalgae to be cultivated in the 1970s and sold as a food supplement. Outdoor farms in Taiwan, Southern Japan and Indonesia produce much of the world supply. Chlorella is also grown in closed indoor production systems.

Unlike spirulina which flourishes in high alkaline high PH water unfriendly to other algae, chlorella grows in normal water conditions. Chlorella is grown in batches, first in indoor tanks, then moved outside to larger ponds. At maximum density the entire batch is harvested by a centrifuge. Chlorella's hard cellulose cell wall protects its nucleus but resists digestion. Farms crack open this cell wall in the drying process, or mechanically crush it.

Today, much chlorella is also grown indoors in photobioreactors-producing a high protein, healthy oil green powder for food, drink and personal care products.

Dunaliella for beta carotene

Dunaliella thrives in water even saltier than the ocean in places like Australia and Israel. Too salty to be eaten as a whole food, its beta carotene is extracted as an oil or powder and sold as a food supplement, antioxidant and color for aquaculture feeds.

Nannochloropsis for protein and omega 3s

Nannochloropsis is being cultivated in outdoor ponds and indoor photobioreactors for nutritional omega-3 oils and proteins for human food and animal and aquaculture feed products.

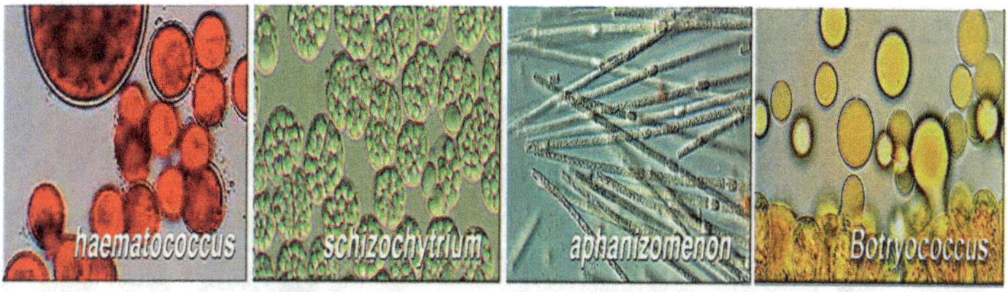

Haematococcus for astaxanthin

Haematococcus is grown in both outdoor ponds and closed systems for astaxanthin, a carotenoid pigment, extracted as a fish feed supplement to color salmon flesh and as a human anti-oxidant food supplement.

Schizochytrium for DHA and EPS

Schizochytrium is a marine microalgae grown in tanks by fermentation, developed as a source of docosahexaenoic acid (DHA), used as a supplement in a wide variety of infant formulas, food and beverages and animal feed products.

Aphanizomenon blue-green algae for nutrition

Aphanizomenon flos-aquae is a nitrogen-fixing blue-green algae. Harvested from Klamath Lake in Oregon, it is sold a food supplement. Wild species grow in lakes and waterways, consuming nutrients are in the water. Some blue-green algae are toxic.

Botryococcus braunii for high oil content

Botryococcus braunii has the highest known lipid or oil content of any microalgae. Because of this much effort has gone into attempts to commercialize botryococcus for biofuel.

Chlamydomonas for alt-meat ingredient

Chlamydomonos is a green single cell algae that swims with two flagella. Can be grown in fermenters and is rich in heme and other meat-like compounds for an alt meat product.

Evolution of agriculture.
1. The Plough • 2. Mechanized Agriculture • 3. Microalgae.

Evolution of food production

Over thousands of years, humans increased food productivity incurring progressively greater environmental costs. About 7000 years ago, irrigation brought water to the land and food surpluses supported the first great river valley civilizations.

A thousand years ago, the invention of an efficient plough in Europe allowed easier soil tilling. Europeans cut down original forests bringing new areas under cultivation and new prosperity. The 19th century industrial revolution introduced mechanized agriculture, climaxing in the 'Green Revolution' exported from the United States in the 1960s and 1970s.

Modern intensive mechanized agriculture has boosted productivity by using plant hybrids and massive fertilizer, pesticide, water and energy inputs. Productivity has been achieved by ignoring hidden costs, such as the consumption of non-renewable fossil fuels, pollution of soil and water through excessive use of chemical fertilizers and the depletion of soils. This ecological damage will be paid by future generations.

Microalgae production is the next evolution in plant productivity, but without the environmental costs.

Algae cultivation is the leap forward

Successful algae cultivation requires an ecological approach to begin with. Spirulina is a living culture and the whole system must be considered. If one factor changes, the entire pond environment changes – quickly. Because algae grows so fast, the result can been seen in hours or days, not seasons like conventional agriculture.

Algae scientists talk of 'balancing pond ecology' for sustainable growth. Pesticides and herbicides would kill many microscopic life forms in a pond, so algae scientists have learned how to balance pond ecology to keep out weed algae and zooplankton algae eaters without pesticides or herbicides.

Ecological food production is the next stage in agriculture. This represents both an increase in productivity and stewardship of the Earth's resources.

The hope of spirulina

A spirulina farm is an environmentally sound green food machine. In shallow ponds, spirulina doubles biomass every 2 to 5 days. This breakthrough yields 20 times more protein than soybeans on the same area, 40 times corn and 400 times beef. Spirulina can flourish in ponds of brackish or alkaline water on unfertile land. It can augment the food supply not by clearing disappearing rainforests, but by cultivating expanding deserts.

In *Spirulina, the Whole Food Revolution,* Larry Switzer wrote: "For the first time since the appearance of man, both wilderness and food productivity can be increased simultaneously with a new technology. This is a choice that man has never had before. The rediscovery of this ancient life as a human food has great implications for us all, now and in the 21st century. It is an example of the myriad of unexpected and astounding solutions to world problems that are now beginning to appear together on this planet."[14]

Chapter 2

A nutrient rich super food for super health

- **Beyond isolated vitamins and minerals**
- **Protein and amino acids**
- **Vitamins and minerals**
- **Essential fatty acids**
- **Phytonutrients and pigments**
- **Comparing green superfoods**

Our modern diet is filled with depleted, over-processed convenience foods. Many people supplement with vitamins and minerals. Now science is looking beyond vitamins to phytonutrients.

Spirulina called a superfood because its nutrient profile is more potent than any other food, plant, grain or herb. The nutrients and phytonutrients in spirulina make it a whole food alternative to isolated vitamin supplements.

Concentrated green super food

Early research documented spirulina's safe consumption by traditional peoples. When scientists discovered spirulina yields 20 times more protein per acre than soybeans, they named it a *food of the future*. Spirulina is 65% protein, higher than any other food with a concentration of vitamins, minerals and other unusual nutrients.

Three to ten grams a day delivers impressive beta carotene, vitamin B-12 and B complex, iron, essential trace minerals and gamma-linolenic acid. Spirulina is rich in phytonutrients that demonstrate a positive effect on health. For undernourished people in the developing world, spirulina brings quick recovery from malnutrition. In Western overfed food culture loaded with unhealthy and depleted foods, spirulina can renourish our bodies and renew our health.

Spirulina is legally approved as a food or food supplement in Europe, Japan and many other countries around the globe. The United States Food and Drug Administration confirmed in 1981 that spirulina can be legally marketed as a food supplement.[1] Many countries have set up food quality and safety standards.

Nutritional depletion of modern foods

Today's food is lower in nutrients than foods produced 50 years ago. Farming practices have depleted soils of minerals. Micro-organisms in the soil contributing valuable mineral content are declining because of overuse of chemical fertilizers. Agribusiness chooses hybrid strains based on harvestability, appearance, and storageability, rather than nutrient content. Long shipping and storage time between harvest and selling reduces nutrient content.

Researchers say increased stress from environmental pollutants and lifestyle demands have increased dietary requirements for certain essential nutrients. As a result, many of us do not trust the quality of our foods. Today, at least some supplements are used by almost everyone.

Beyond isolated vitamins and minerals

Vitamins and minerals in foods are bound to natural food complexes with proteins, carbohydrates and lipids. The human body recognizes this entire food complex as food. Many supplements are combinations of isolated USP vitamins and minerals. These are formulated to claim 100% of the Daily Value (DV) on labels. But if these vitamins and minerals are not bound to anything, they may have a different chemical structure than those found in foods.

Formulas may ignore antagonistic and synergistic effects of vitamins and minerals both in regard to absorption and metabolic reactions. Complex factors in whole foods that aid absorption may be missing in laboratory formulated vitamins and minerals. It is well known that many supplements, especially calcium and iron, are not well absorbed.

Supplement megadoses attempt to overcome absorption problems with an approach that more is better. But absorption of vitamins and minerals may be limited by uptake mechanisms in the intestines, and megadoses largely excreted.

Technicians formulated chemical chelators, transporters and time release agents to make mineral supplements better absorbed. Vitamins and minerals have been extracted from food sources. Food-form type supplements recombine USP vitamins and minerals in a vat with yeast bacteria. It is claimed these products are 'just like food'.

Most people believe it is better to get nutrients from natural foods. Since many conventional foods are nutrient depleted, more people take spirulina and green superfoods. These whole foods offer functional nutrients and phytochemicals, new frontiers for disease prevention research, beyond isolated vitamin and mineral supplements.

Protein and amino acids

The building blocks of life are protein and amino acids. When comparing proteins, several criteria should be considered: protein quantity, amino acid quality, usable protein, digestibility, and side effects from fat, calorie and cholesterol.

Spirulina has the highest protein of any natural food (65%). More than animal and fish meat (15-25%), soybeans (35%), dried milk (35%), peanuts (25%), eggs (12%), grains (8-14%) or milk (3%).

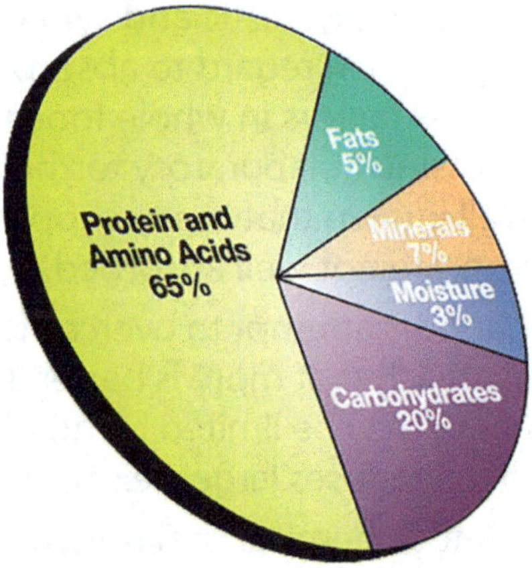

Composition of spirulina.

Protein is composed of amino acids. Essential amino acids cannot be manufactured in the body and must be supplied in the diet. Non-essential amino acids are needed too, but the body can synthesize them. A protein is considered complete if it has all the essential amino acids. Spirulina is just that, a complete protein.

The body requires amino acids in specific proportions. If a food is low in one or more, those amino acids are called limiting, and the body cannot use all amino acids completely. The ideal proportion of amino acids is eggs. All other foods have some limiting amino acids.

Spirulina amino acid composition[a]

Essential Amino Acids	per 10 grams	% total
Isoleucine	350 mg	5.6 %
Leucine	540 mg	8.7 %
Lysine	290 mg	4.7 %
Methionine	140 mg	2.3 %
Phenylalanine	280 mg	4.5 %
Threonine	320 mg	5.2 %
Tryptophan	90 mg	1.5 %
Valine	400 mg	6.5 %
Non-Essential Amino Acids	**per 10 grams**	**% total**
Alanine	470 mg	7.6 %
Arginine	430 mg	6.9 %
Aspartic Acid	610 mg	9.8 %
Cystine	60 mg	1.0 %
Glutamic Acid	910 mg	14.6 %
Glycine	320 mg	5.2 %
Histidine	100 mg	1.6 %
Proline	270 mg	4.3 %
Serine	320 mg	5.2 %
Tyrosine	300 mg	4.8 %
Total Amino Acids	**6200 mg**	**100.0 %**

a. Earthrise Farms, 1995.

Limiting amino acids in spirulina are methionine and cystine, but it is still higher in these amino acids than grains, seeds, vegetables and legumes, and higher in lysine than all vegetables except legumes. Spirulina complements vegetable protein and increases the amino acid quality. Over 100% of the daily essential amino acid requirements for a typical adult male are supplied by using only 36 grams of spirulina, about 4 heaping tablespoons.

Net protein utilization and usable protein

Feeding tests rank proteins by Net Protein Utilization (NPU) value. Dried eggs (94) have the highest value, followed by milk (70-82), fish (80) and meat (67). Spirulina (62) is similar to grains and has a higher NPU than nuts).

Adult essential amino acid (EAA) requirements provided by spirulina[a]

EAA	requirement g / day	spirulina gm / 10g	spirulina %/ 10g
Isoleucine	0.84	0.35	42 %
Leucine	1.12	0.54	48 %
Lysine	0.84	0.29	35 %
Methionine[b]	0.70	0.20	29 %
Phenylalanine[c]	1.12	0.58	52 %
Threonine	0.56	0.32	43 %
Valine	0.98	0.40	41 %

a. Jassby 1983, FNB 1975, Earthrise Farms 1995.
b. includes cystine. c. includes tyrosine.

Protein quantity and quality for spirulina and other protein food sources[a]

Food	Protein %	NPU %	Usable Protein %
Spirulina[b]	65	62	40
Dried eggs, whole	47	94	44
Brewers yeast	45	50	23
Soy flour, whole	37	61	23
Dried milk, skim	36	82	30
Cheese, parmesan	36	70	25
Wheat germ	27	67	18
Peanuts	26	38	10
Chicken[c]	24	67	16
Fish[c]	22	80	18
Beef[c]	22	67	15
Sesame seed	19	60	11
Oats, whole flour	15	66	10
Wheat, whole flour[c]	14	63	9
Tofu, moist	8	65	5
Brown rice	8	60	5

a. Switzer, The Whole Food Revolution, 1982, pg 21.
b. value for spirulina NPU from O. Ciferri, Spirulina.
c. values are for highest protein levels for these groups.

By multiplying protein quantity by the NPU, we determine usable protein as a percentage of the food's composition. Spirulina is second only to dried eggs, and higher than any common foods in the form they are usually purchased.

Protein digestibility is important for many people

Spirulina has no cellulose in its cell walls. Composed of soft mucopolysaccharides makes it easily assimilated - 85 to 95% digestible. This is important for people with intestinal malabsorption. Many older people have difficulty digesting complex proteins, and are on restricted diets. They find spirulina protein easy to digest. Spirulina is effective for victims of malnutrition diseases like kwashiorkor, where intestinal absorption has been damaged. Given to malnourished children, it is more effective than milk powders because milk lactic acid can be hard to absorb.

'Side effects' - fat, calories and cholesterol

Spirulina's fat content is 5%, far lower than almost all other protein sources. Ten grams has only 36 calories and virtually no cholesterol. This means spirulina is a low-fat, low-calorie, cholesterol-free source of protein, and is not loaded with the fat, grease, calories and cholesterol of meat and dairy protein.

One tablespoon (10 grams) of spirulina contains only 1.3 mg of cholesterol and 36 calories. In contrast, a large egg yields about 300 mg of cholesterol and 80 calories, while providing only the same protein as the tablespoon of spirulina.

People in developed countries usually consume more than enough protein along with excessive fat, calories and cholesterol. Thus, less dairy and meat protein are recommended.

Some people need higher protein intake, but not so many calories. So additional protein must be low in calories to avoid an unnecessary weight increase. Spirulina is suitable in these cases where the calorie cost is far lower than from dairy, meat and fish.

Spirulina vitamin content

Vitamins[a]	per 10 grams	U.S. DV	% DV
Vitamin A (beta carotene)	23000 IU	5000 IU	460 %
Vitamin C	0 mg	60 mg	0 %
Vitamin E (a-tocopherol)	1.0 IU	30 IU	3 %
Vitamin K	200 mcg	80 mcg	250 %
Vitamin Bl (thiamin)	0.35 mg	1.5 mg	23 %
Vitamin B2 (riboflavin)	0.40 mg	1.7 mg	23 %
Vitamin B3 (niacin)	1.40 mg	20 mg	7 %
Vitamin B6 (pyridoxine)	80 mcg	2 mg	4 %
Folate (folic acid)	1 mcg	0.4 mg	0 %
Vitamin B12 (cyanocobalamin)	20 mcg	6 mcg	330 %
Biotin	0.5 mcg	0.3 mg	0 %
Panthothenic Acid	10 mcg	10 mg	1 %
Inositol	6.4 mg	***	***

a. Earthrise Farms 1995.

Vitamins - protectors of health

A ten gram spirulina serving (20 tablets, or 1/3 ounce) supplies a rich profile of vitamins we need.

Natural beta carotene (provitamin A)

Spirulina is the richest beta carotene food, ten times more concentrated than carrots. Ten grams provide 23,000 IU (14 mg) of beta carotene, 460% of the U.S. Daily Value (DV) of Vitamin A. High doses of Vitamin A may be toxic, but beta carotene in spirulina and vegetables is safe, because humans convert beta carotene to Vitamin A only as needed. Vitamin A is important in maintaining mucous membranes and pigments necessary for vision. Vitamin A deficiency is one of the worst malnutrition diseases in the developing world.

Beta carotene has therapeutic effects, reducing serum cholesterol and cancer risks. Published studies show beta carotene reduces the risks of all kinds of cancers, including lung, throat, stomach, colon, gastrointestinal tract, breast and cervix.

Best beta carotene vegetables[a]

Food	serving size	IU of beta carotene
spirulina[b]	**1 heaping tbsp. (10 g)**	**23000**
papaya	1/2 medium	8867
sweet potato	1/2 cup, cooked	8500
collard greens	1/2 cup, cooked	7917
carrots	1/2 cup, cooked	7250
chard	1/2 cup, cooked	6042
beet greens	1/2 cup, cooked	6042
spinach	1/2 cup, cooked	6000
cantaloupe	1/4 medium	5667
chlorella[c]	**50 tablets (10 g)**	**5000**
broccoli	1/2 cup, cooked	3229
butternut squash	1/2 cup, cooked	1333
watermelon	1 cup	1173
peach	1 large	1042
apricot	1 medium	892

a. Vegetarian Times, "Recipes with A+ Nutrition", May 1986, pg 47.
b. Earthrise Farms, 1995. c. Yaeyama Chlorella, 1995.

Cancer seems to be partly caused by environmental factors, and scientists say risks can be reduced by increasing protective factors, especially beta carotene, in the diet. Natural beta carotene is a combination of *cis* and *trans* isomers, whereas synthetic beta carotene has only the *trans*, with lower antioxidant potential.

Although beta carotene is best known, spirulina has an antioxidant rich complex of ten carotenoids. These mixed carotenes and xanthophylls function at different sites in the body and work synergistically with the other vitamins, minerals and phytonutrients. This is more effective than a synthetic beta carotene supplement.

Even if you don't eat the recommended 4 to 9 servings of fruits and vegetables every day (most people eat only 1 to 2 including french fries), get your natural carotenoid antioxidant protection from spirulina tablets or powder every day.

Vitamin B-12 and B-complex vitamins

Spirulina is the richest source of B-12, higher than beef liver, chlorella or sea vegetables. B-12 is necessary for development of red blood cells, especially in bone marrow and nervous system. Although primary B-12 deficiencies, pernicious anemia and nerve degeneration, are quite rare, because B-12 is the most difficult vitamin to get from plant sources, vegetarians have taken to spirulina.

Ten grams contain 20 mcg of Vitamin B-12, 330% DV, using the microbiological assay. Vitamin B-12 label content claims for foods and dietary supplements are based on the approved microbiological assay. This method is used for spirulina, because it is being compared with the B-12 content of other foods and vitamins.

An alternative method developed in the 1980s, radio assay, has measured the B-12 assumed to be bioavailable to humans. Radio assay found higher levels of B-12 analogs and lower levels of bioavailable B-12 in all foods and supplements, and shows spirulina has only 20% of the original B-12. Even using these lower levels, it is the best non-animal source of Vitamin B-12.

Some incomplete research has suggested B-12 analogs could block B-12 absorption, based on limited results with very few individuals, and did not consider B-12 non-absorption due to folic acid or other dietary deficiency. In nearly 40 years, there have been no complaints of a vitamin B-12 deficiency from spirulina consumers, including children and vegetarians.

One tablespoon provides significant quantities of thiamin (23% DV), required for functioning of nerve tissues, riboflavin (23% DV), to gain energy from carbohydrates and proteins, and niacin (7% DV) for healthy tissue cells. Spirulina is a richer source of these vitamins than common whole grains, fruits and vegetables and some seeds. Other B vitamins, B-6, niacin, biotin, pantothenic acid, folic acid, inositol and Vitamin E are also present in smaller amounts.

Spirulina mineral content

Minerals[a]	per 10 grams	U.S. DV	% U.S. DV
Calcium	70 mg	1000 mg	7 %
Iron	10 mg	18 mg	55 %
Phosphorus	80 mg	1000 mg	8 %
Magnesium	40 mg	400 mg	10 %
Zinc	300mcg	15 mg	2 %
Selenium	10mcg	70 mcg	14 %
Copper	120mcg	2 mg	6 %
Manganese	500mcg	2 mg	25 %
Chromium	25mcg	120 mcg	21 %
Sodium	90 mg	2400 mg	4 %
Potassium	140 mg	3500 mg	4 %
Germanium	60mcg	-	-

a. Earthrise Farms, 1995.

Naturally colloidal minerals

Algae absorbs many trace elements while growing and these minerals are well assimilated by the human body. Mineral content varies depending on where it is grown and minerals in the water.

The best natural iron supplement

Iron is the most common mineral deficiency worldwide, especially for women, children and older people. Women on weight loss diets typically do not get enough iron, and can become anemic. Iron is essential for strong red blood cells and a healthy immune system. Spirulina is a rich iron food, 10 times higher than common iron foods. Ten grams supply up to 10 mg of iron, 55% of the DV.

Spirulina iron is easily absorbed by the human body. It is theorized that its blue pigment, phycocyanin, forms soluble complexes with iron and other minerals during digestion, making iron more bioavailable. Hence, iron in spirulina is over twice as absorbable as the form of iron found in vegetables and most meats.[2]

Typical iron supplements are not well absorbed. Studies show iron in spirulina is 60% better absorbed than supplements such as iron sulfate. For all people who need iron supplements, spirulina is one of the best sources.

Best food sources of Iron[a]

Food	serving size	mg Iron
Spirulina[b]	**1 tbsp. (10g)**	**10.0**
Chlorella[c]	**1 tbsp. (10g)**	**10.0**
Chicken liver, cooked	3 ounces	7.2
Crab, pieces, steamed	1/2 cup	6.0
Beef liver, fried	1/2 cup	5.3
Soybeans, boiled	1/2 cup	4.4
Blackstrap molasses	1 tbsp.	3.2
Spinach, cooked	1/2 cup	3.2
Beef, sirloin, broiled	3 ounces	2.9
Potato, baked	one	2.8
Scallops, steamed	3 ounces	2.5
Pistachios, dried	1/4 cup	2.2
Broccoli, cooked	1 spear	2.1
Cashews, dry-roasted	1/4 cup	2.1
Turkey, dark meat	3 ounces	2.0
Spinach, raw chopped	1/2 cup	0.8

a. The Complete Book of Vitamins and Minerals for Health, pg. 182.
b. Earthrise Farms, 1995. c. Yaeyama Chlorella, 1995.

Calcium, magnesium, zinc and trace minerals

Spirulina supplies, gram for gram, more calcium than milk. Ten grams supply 7% DV. Calcium is important for bones and neural transmissions to the muscles. Deficiencies lead to osteoporosis in older women. Ten grams supply 10% DV for magnesium, facilitating absorption and regulating blood pressure. Spirulina is low in iodine and sodium, and no problem for salt-restricted diets.

Humans need essential trace minerals for the functioning of enzyme systems and other physiological functions. Deficiency of trace minerals are widespread. Ten grams supply manganese (25% DV), chromium (21% DV), selenium (14% DV), copper (6% DV) and zinc (2% DV).

Essential fatty acids

Essential fatty acids (EFA) promote cholesterol normalization and are precursors for hormones called prostaglandins. Spirulina has 4 to 7% lipids, or fats, and most are essential fatty acids. Ten grams have 225 mg as linoleic and gamma-linolenic acid (GLA).

GLA is the precursor to the body's prostaglandins – master hormones that control many functions. Dietary saturated fats and alcohol can cause in GLA deficiency and suppressed prostaglandins. GLA deficiency figures in many diseases and health problems, so a food source of GLA can be important.

The only other known sources of dietary GLA are mother's milk and extracts of evening primrose, black currant and borage seeds. Spirulina has concentrated GLA, and a 10 gram serving has 135 mg. 500 mg of evening primrose oil has 45 mg. GLA is about 20 to 25% of the lipids of spirulina, compared to 9% for evening primrose oil.

Spirulina essential fatty acids[a]

		mg per 10 grams	% total
C 14:0	Myristic	1 mg	0.2 %
C 16:0	Palmitic	244 mg	45.0 %
C 16:1	Palmitoleic	33 mg	5.6 %
C 17:0	Heptadecanoic	2 mg	0.3 %
C 18:0	Stearic	8 mg	1.4 %
C 18:1	Oleic	12 mg	2.2 %
C 18:2	Linoleic	97 mg	17.9 %
C 18:3	**Gamma-linolenic**	**135 mg**	**24.9 %**
C 20	Others	14 mg	2.5 %
Total		**546 mg**	**100 %**

a. Earthrise Farms 1995.

Dietary sources of GLA

Food sources	Oil extracts
Mother's milk	Evening primrose plant
Spirulina	Black currant and borage seeds

Phytonutrients and Pigments

These functional nutrients have no Recommended Daily Value, but are known to benefit health. They include pigments, polysaccharides and glycolipids. Pigments help synthesize enzymes necessary for regulating body metabolism. Spirulina's dark color comes from its pigments which harvest different wave lengths of sunlight.

Phycocyanin (algae-blue)

The most important pigment in spirulina, this protein complex is about 10-15% of the entire weight. Phycocyanin evolved a billion years before chlorophyll and may be the precursor to chlorophyll and hemoglobin. It has both magnesium and iron in its molecular formation, and phycocyanin may be the origin of life common to both plants and animals.[3] Research indicates it stimulates the immune system.

Chlorophyll (nature's green magic)

Green foods have high chlorophyll, known for cleansing and detoxifying, sometimes called 'green blood' because it looks like the hemoglobin molecule in human blood. Chlorophyll has a magnesium ion, giving it a green color, and hemoglobin has iron, giving it a red color.[9] Spirulina's beneficial effect on anemia could be due to this similarity of chlorophyll and hemoglobin and its bioavailable iron. Spirulina has 1% chlorophyll, one of nature's highest with high chlorophyll-a. Chlorella has 2 to 3%, mostly chlorophyll-b.

Carotenoids (natural antioxidants)

About half of the yellow/orange pigments in spirulina are carotenes: Alpha, Beta and Gamma. About half are xanthophylls: Myxoxanthophyll, Zeaxanthin, Cryptoxanthin, Echinenone, Fucoxanthin, Violaxanthin and Astaxanthin. Mixed carotenoids make up 0.37% of spirulina. Although beta carotene is best known, this mixed carotenoid complex functions at different sites in the body and works synergistically to enhance antioxidant protection.

Spirulina natural pigments

Pigments[a]	Color	per 10 grams	% total
Phycocyanin	**(blue)**	**1400 mg**	**14 %**
Chlorophyll	**(green)**	**100 mg**	**1.0 %**
Carotenoids	**(orange)**	**37 mg**	**0.37 %**
Carotenes	***54 %***	***20 mg***	***0.20 %***
Beta carotene	45 %	17 mg	0.17 %
Other Carotenes	9 %	3 mg	0.03 %
Xanthophylls	***46 %***	***17 mg***	***0.17 %***
Myxoxanthophyll	19 %	7 mg	0.07 %
Zeaxanthin	16 %	6 mg	0.06 %
Cryptoxanthin	3 %	1 mg	0.01 %
Echinenone	2 %	1 mg	0.01 %
Other Xanthophylls	6 %	2 mg	0.02 %

a. Earthrise Farms 1995.

Polysaccharides

Spirulina contains 15 to 25% carbohydrate and sugar. The primary forms are rhamnose and glycogen, two polysaccharides which are easily absorbed by the body. Spirulina offers quick energy, without taxing the pancreas or precipitating hypoglycemia.

Glycolipids and Sulfolipids

Sulfolipids in blue-green algae are 'remarkably active' against the AIDS virus. Three classes of lipids in spirulina are called neutral lipids, glycolipids and phospholipids. Glycolipids are 40% of the lipids, and contain sulfolipids from 2-5% of the total lipids.[4]

Enzymes

Enzymes are catalysts for chemical changes. Dried spirulina contains a number of enzymes. Superoxide dismutase (SOD), is important in quenching free radicals and in retarding aging. SOD enzyme activity ranging from 10,000 to 37,500 units per ten grams has been found in spirulina powder.

Comparing green superfoods

Today more people understand the need for green vegetables. At the same time, there is concern with the quality of foods and vegetables grown on mineral depleted soils.

Green superfoods go beyond green vegetables because they are packed with beneficial nutrients. They go beyond isolated vitamin and mineral supplements, because they are rich in functional nutrients and phytonutrients. Research reports link phytonutrient, antioxidant and protective substances in plant foods with prevention of degenerative diseases.

Nutrient dense green superfoods are often consumed as tablets or capsules or by mixing powder in drinks. These fast foods pick up your energy, especially if you do not have time to eat the recommended 4 to 9 servings of fresh fruits and vegetables every day.

Green superfood supplements have become more popular. Spirulina and chlorella are cultivated algae. Aphanizomenon flos-aquae (referred to as 'blue-green algae') is harvested from a lake in Oregon. Barley grass and wheat grass are two superfoods from cultivated young cereal plants, harvested before they become grains. All five chlorophyll-rich foods are specially harvested to maximize purity, potency and quality.

Comparing nutrients, data provided by manufacturers is shown on the next page for a 10 gram serving.

Algae and grasses are the foundation of life on Earth, harvesting sunlight. Their deep green color glows with the vitality from the rainbow of natural pigments which power, protect and cleanse them while they grow. These natural foods will nourish, energize and cleanse your body naturally. Eating just a little of these concentrated green foods every day will benefit your health. Eating lower on the food chain will benefit the health of our planet.

Green superfood nutrient comparison

Composition	spirulina[a] algae	chlorella[b] algae	aphaniz.[c] algae	barley[d] grass	wheat[d] grass
Protein	62 %	60 %	58 %	25 %	25 %
Carbohydrates	19 %	18 %	25 %	54 %	54 %
Fats (lipids)	5 %	10 %	5 %	4 %	4 %
Minerals (ash)	9 %	7 %	7 %	12 %	12 %
Moisture	5 %	5 %	5 %	5 %	5 %
Vitamins (per 10 grams)					
Beta carotene	23000 IU	5000 IU	12000 IU	5000 IU	5000 IU
Vitamin C	0 mg	4 mg	6 mg	31 mg	31 mg
Vitamin E	1 IU	1.5 IU	1.3 IU	3 IU	3 IU
Thiamin, B1	0.35 mg	0.17 mg	0.05 mg	0.03 mg	0.03 mg
Riboflavin, B2	0.40 mg	0.50 mg	0.50 mg	0.20 mg	0.20 mg
Niacin, B3	1.40 mg	2.80 mg	1.30 mg	0.75 mg	0.75 mg
Vitamin B6	80 mcg	140 mcg	110 mcg	128 mcg	128 mcg
Vitamin B12	20 mcg	5 mcg	32 mcg	3 mcg	3 mcg
Folacin	1 mcg	*	10 mcg	108 mcg	108 mcg
Biotin	0.5 mcg	*	3 mcg	11 mcg	11 mcg
Pantothenic acid	10 mcg	*	60 mcg	240 mcg	240 mcg
Inositol	6 mg	*	*	*	*
Minerals (per 10 grams)					
Calcium	70 mg	30 mg	140 mg	52 mg	52 mg
Iron	10 mg	10 mg	3.5 mg	6 mg	6 mg
Magnesium	40 mg	30 mg	22 mg	10 mg	10 mg
Sodium	90 mg	36 mg	27 mg	3 mg	3 mg
Potassium	140 mg	80 mg	120 mg	320 mg	320 mg
Phosphorus	90 mg	90 mg	50 mg	52 mg	52 mg
Zinc	0.3 mg	1.2 mg	0.2 mg	0.5 mg	0.5 mg
Manganese	0.5 mg	*	0.3 mg	1.0 mg	1.0 mg
Copper	120 mcg	*	40 mcg	200 mcg	200 mcg
Chromium	25 mcg	*	5 mcg	*	*
Phytonutrients (per 10 grams)					
Phycocyanin	1400 mg	none	*	none	none
Chlorophyll	100 mg	280 mg	200 mg	55 mg	55 mg
Total Carotenoids	37 mg	*	*	*	*
Gamma Linolenic Acid	135 mg	*	*	none	none
Glycolipids	200 mg	*	*	*	*
Sulfolipids	10 mg	*	*	*	*

a. Earthrise Farms, 1995. b. Yaeyama Chlorella, 1995.
c. Cell Tech, Alpha Sun. d. Cereal Grass, ed. by Ronald Seibold. * no data available.

Chapter 3

Enjoy clean green energy for vibrant health

- **Why does spirulina benefit your health?**
- **How to use spirulina powder, tablets, capsules**
- **Innovative fresh, frozen and dry crunchies**
- **Eat lighter for natural wight control**
- **Lower risk of cancer, lower cholesterol**
- **Popular for detox, fasting and cleansing**
- **Endurance for athletes and bodybuilders**
- **Great for children and mothers**

Spirulina has proven health benefits and is most effective when used with a natural food diet in a personal strategy for self-care. People who take spirulina each day report cumulative health benefits, for energy and stamina, detox and cleansing, or maintaining your natural weight. Children love green spirulina drinks!

Why does spirulina benefit your heath?

Proven health benefits from 40 years of customer experience and published international medical research. Six major benefits from the phycocyanin, polysaccharides, antioxidants and phytonutrients in a 3 gram serving per day:

Enhances immune system protection. Promotes macrophage function, T-cell proliferation, anti-viral and natural killer cell activity. Important in the regulation of antibody production.

Promotes beneficial intestinal flora. Supports lactobacillus in the intestines for three benefits: better digestion and absorption, protection from infection, and stimulation of the immune system.

Supports cardiovascular heath. Protects from oxidative stress. Spirulina may lower total cholesterol, LDL cholesterol and triglycerides, while raising HDL good cholesterol.

Strengthens brain health and anti-aging. Spirulina may prevent memory loss by lessening protein accumulation, reducing oxidative damage and raising catalase activity.

Detoxifies pollutants, heavy metals, radioactive compounds. Health practitioners recommend detoxifying from lifetime of accumulated toxins. Spirulina phytonutrients phycocyanin and chloroplyll bind accelerate excretion.

Improves eye and cell health. Beta-carotene, amino acids and zinc may defer age-related vision problems such as macular degeneration and cataracts. Antioxidant and anti-inflammatory action protects cells from damaging oxygen radicals.

Spirulina powder, tablets and capsules.

How to use spirulina powder

Spirulina powder has a dark green or blue-green color. Your body feels energy within minutes because powder is naturally digestible. It provides quick energy and nourishment between meals or in place of a meal. Some have asked whether you can take too much. It is a perfectly safe natural food. Some people take a tablespoon or more each day.

Author with a morning smoothie in a blender (1994).

Add to fruit or vegetable juice in a blender. Start with one teaspoon (5 gm). Later on increase. Many use one tablespoon (10 gm) per drink.

Don't stick a wet spoon in a bottle or put a spoonful directly into liquid. Water will stick the powder on your spoon. Add slowly while stirring.

Spirulina will keep well if handled properly. Dry powder absorbs water from the air. Keep bottle sealed when not in use. You don't need to refrigerate, but keep in a dry, dark place.

Tasty in soups, salads, pasta, breads and taboulie. Even a little will give food a dark green color. With recipes that require cooking, heat spirulina as little as possible.

Morning Smoothie
Instant breakfast to start your day

Blend one tablespoon spirulina powder in 2 cups of tropical blend juice (or orange, apple or pineapple juice). Options: add one whole fruit (banana, orange or peach), almonds, sunflower seeds or flavors such as vanilla or lime to suit your taste. (Mix well in a blender. Makes 2 servings).

Veggie Cocktail
Mid afternoon vegetable pick up

Blend one tablespoon spirulina powder in 2 cups of vegetable juice. Options: add whole vegetables, herbs (parsley, dill weed) or spices (cayenne, horseradish) to suit your taste. (Mix well in a blender. Makes 2 servings).

Convenient tablets and capsules

Tablets can be made without sugar, starch, fillers, preservatives, stabilizers, colors, coatings, and with only a minimum of tableting agents. Tablet color should be a uniform dark green without light colored specks. Capsules should be free of fillers or additives, and vegetable capsules are available.

Bottles often suggest a 6 tablet serving (3 grams), but you can take more- 10, 15 or 20 a day. Twenty 500 mg tablets are equal to a heaping tablespoon of powder (10 grams). Tablets have the same benefits as powder, but digestion takes about an hour. Because it is a whole food, you can take tablets between meals.

If you are using spirulina to balance your diet and help eat lighter meals, take tablets or capsules an hour before you eat. During the day when your energy runs low, take tablets and see how your body feels one or two hours later. Both tablets and capsules are helpful with water after coffee or alcohol.

Spirulina crunchies, fresh paste and frozen portions.

Innovative forms of pure spirulina

Most available spirulina products use powder produced by large commercial farms around the world for tablets and capsules. Yet, over this last decade, many smaller spirulina microfarms have emerged in Asia, Africa, Europe and the Americas. These local and regional growers have introduced new forms of spirulina beyond dry powder.

SpiruSource: frozen, fresh, dry crunchies. 2019.

Fresh, frozen and dry crunchy noodles

If you are fortunate to live near a spirulina microfarm, you may be able to get fresh spirulina harvest, fresh frozen spirulina or low temperature dried noodles or crunchies. Purchase online, at farmers markets or select retail stores.

Mixing fresh spirulina in a blender for a fruit smoothie.

Fresh harvest has almost no taste

Because it grows in alkaline water at high 10-11 pH, when the culture is properly managed, fresh harvest spirulina is safe to eat. Harvesting with microscreens, it has a consistancy from yogurt to tofu depending on water content. It has almost no taste. It should be immediately refrigerated, then this high protein food can last for 3-7 days before spoiling. Best to consume right away.

Fresh spirulina drinks and dips

Denise Fox and Tarome prepare fresh spirulina savory and sweet aquamole dips. (France 2011)

Fresh frozen: Drop into juice, let melt, stir and drink.

Convenient individual frozen portions

Fresh harvest can be frozen to preserve it for a year or longer. Frozen into individual 1/2 ounce (14g) servings, in packets or cups, a serving can be dropped into a drink, melted and stirred for a fresh green drink.

Crunchies and noodles: Sprinkle on foods or snacks.

Tasty crunchy noodles

At spirulina microfarms fresh harvest is dried on trays in fruit and vegetable dehydrators over several hours. These lower temperature dried noodle pieces are so mild with a tasty savory flavor, they are good sprinkled on other foods or munched on directly as snacks.

Green food drink mixes with spirulina and other superfoods are popular supplements.

Green yourself for vibrant health

Spirulina is effective as part of a lighter, fresher, more natural diet advocated by leading medical experts for enjoying a healthier, energized and longer life.

Reduce processed convenience foods

Many people rely on convenience foods rich in fats, carbos and sugar, and low in vegetables and fiber. These foods can raise body weight. cholesterol and worsen digestive problems. A body overloaded with non-nutritious foods doesn't assimilate enough quality nutrients, starving for more nutrients, triggering appetite and compulsive overeating. Without exercise, extra calories stay on as fat. People remain trapped in a difficult cycle. Taking nutrient dense spirulina daily can help reduce the desire for convenience foods.

Spirulina products in the USA.

Natural weight control

A more natural diet satisfies hunger because it satisfies the body's real hunger for nutrition. Spirulina is a concentrated natural food and can help restore natural body weight. Many people use it along with a low carbohydrate diet and exercise to eat lighter meals and avoid fattening snacks.

Take a heaping teaspoon of powder (about 5 grams), or at least 6 tablets one hour before meals or snack breaks. You know when you're going to be hungry, so plan ahead. This superfood can help satisfy appetite. It is not an appetite suppressant, and contains no drugs or chemicals that trick the body. It is concentrated, easily digested and rich in iron, often deficient in people on low cal diets.

Slow but steady weight loss is desirable. Spirulina helps us to lighten up, and provides the energy to make the switch from a unhealthy diet to lighter, more nutrient rich foods.

Reducing pre-menstrual syndrome (PMS)

Studies show women with more severe PMS have unusually low levels of certain nutrients, so many health experts urge a nutritional approach including B-complex, magnesium, zinc, beta carotene and GLA. Spirulina contains many of these nutrients and several PMS supplements contain spirulina.

Reducing risks of cancer

Scientific studies have clearly shown eating foods rich in beta carotene will lower risks of all kinds of cancer. Eat two servings each day of beta carotene vegetables such as carrots, sweet potatoes, greens, squashes, and fruits such as papaya and cantaloupe.

Fortunately, spirulina is over ten times more concentrated in beta carotene than any other vegetable. Only three grams daily offers 140% of the US RDA of Vitamin A in the safe form of natural beta carotene. Plus, its entire carotenoid complex offers antioxidant protection at different sites in the body.

Lowering cholesterol

Studies show as cholesterol levels decline, risks of heart attacks and strokes decline. Over 70% of Americans understand the need to reduce cholesterol-rich foods that increase cholesterol which creates plaque and blocks the arteries, causing atherosclerosis, leading to heart disease. Spirulina is a helpful for cholesterol reduction. Research in Japan with male volunteers shows only 4 grams of spirulina a day (about 8 tablets) significantly reduced cholesterol.

Nutrient dense anti-aging food

Many older people have restricted diets or suffer from poor digestion and low energy. Concentrated foods like spirulina are excellent choices: 60% protein and easy-to-digest. Vegetable low-fat protein means seniors can lighten up on meat centered diets that aggravate arthritis and raise cholesterol.

Protective anti-oxidant beta carotene is good for eyes and vision. Spirulina builds healthy lactobacillus, aids assimilation and elimination. Phycocyanin assists detoxification. More older people desire less meat, so protein sources like spirulina are helpful. B-12 absorption decreases with age, so B-12 rich spirulina is a good choice. Spirulina contains the essential fatty acid GLA for healthy skin.

Many detox and cleansing products have spirulina.

Promotes detox, fasting and cleansing

Going without solid food for three days to a week or longer allows the body to cleanse and renew. People who benefit from fasting report being physically stronger and psychologically clearer. Spirulina is easy to digest and provides energy and stamina. Because it is a nutrient dense food, spirulina makes fasting easier.

Fasting requires discipline and should be done for limited periods. When fasting is over, phase in light simple meals slowly over several days before moving back to a normal diet. The first few days of fasting may result in discomfort due to detoxification. If side effects are severe, fasting should stop.

Assists colon cleansing

Wastes accumulate in pockets in the colon. Never removed, this leads to constipation, weakened digestion, poor nutrient absorption, and in later life, colon cancer. Older people may suffer from chronic constipation, and need laxatives for symptomatic relief. Spirulina is a cleansing food that empties toxins out of the body tissues into the lymph and burns up mucoforming substances.

Helps recovery the morning after

Ten or more spirulina tablets after an evening on the town can reduce the worst effects of hangovers. With two glasses of water for rehydration, spirulina helps rebuild depleted nutrients. It does not aid sobriety, but helps avoid a depleted feeling.

Protein, energy and body building powders often contain spirulina and green superfoods.

Energy for athletes and bodybuilders

Taken before competition, athletes say spirulina delivers energy, improves stamina, increases endurance of runners. Backpackers, cyclists and climbers take tablets for more energy and stamina per weight than conventional foods. This high intensity food is perfectly suited for high intensity training.

Proteins are essential for endurance training and to regenerate body tissue. For the bodybuilder, spirulina offers 65% protein, easy to digest and low in fat. Bodybuilders take 10 grams up to three times a day. Before competition it gives an energy boost with sustaining power, mixed with milk, egg, honey and juice. Taken before meals it can satisfy appetite for maintaining competitive weight.

World Class athletes use spirulina to improve performance. At the largest Chinese training center for 2000 athletes, trainers reported it improved recovery and boosted the immune system, allowing athletes to intensify training. Track stars have used spirulina for many years to create and mend muscle mass. Marathoners consume spirulina for endurance and to ward off cramping.

Spirulina, chlorella and marine algae extracts are in many personal care products.

Beautiful skin and personal care

For more beautiful skin and hair, spirulina and other algae extracts are key ingredients in many personal care products such as scrubs, masks, creams, shampoos and cleansers. Skin creams containing spirulina could be beneficial for skin conditions due to spirulina antioxidant, antimicrobial and anti-inflammatory effects and its ability to eliminate toxins in the body and skin.

Kobe West enjoys a fresh spirulina green drink. (California 2021)

Safe and beneficial for children

Children benefit the most from spirulina. It is highly nutritious plant protein and easily digested. Even undernourished children with poor nutrient absorption can assinimlate spirulina and recover from malnutrition. Spirulina builds tissue, improves vision and concentration, strengthens the immune system, enhances resistance to infection and the ability to heal. Kids love spirulina green drinks!

Good for pregnant and nursing mothers

Nutrition during pregnancy is important. Iron deficiency anemia in mothers and children is a prevalent nutritional disorder.[17] Pregnant women need spirulina's extra easy-to-digest protein and bioavailable iron. In India and Vietnam, it is prescribed for pregnant and nursing mothers.

Green superfoods are perfect for food storage.

Survive and thrive with superfood reserves

Millions of people store several years of food for security. Their concerns are earthquakes, environmental catastrophe, nuclear accidents, food supply interruption or economic collapse. Spirulina has always been popular with people who store food because it is concentrated, lightweight and portable- perfect for an emergency food supply.

Spirulina nutrition can turn a storage food diet from merely surviving to really thriving. Much less space is needed compared to bulky grains. Ideal for urban or suburban dwellers.

For long term storage, keep in airtight and watertight containers, without exposure to light or excessive heat. Rotate these foods and replace with fresh stocks. Regular plastic jars may not protect against oxygen penetration which can destroy beta carotene over time. Oxygen barrier plastic jars allow storage for years.

Enjoy this green superfood every day

People who enjoy spirulina each day report the most health benefits, whether for energy, cleansing or dieting. One benefit you may notice from regular use is more beautiful skin. We need all the protection we can get from the right foods.

Chapter 4

New research reveals health benefits

- **Increase anti-viral activity**
- **Lower cholesterol**
- **Anti-aging and neuroprotective**
- **Reduce risk of cancer**
- **Stimulate the immune system**
- **Reduce kidney toxicity**
- **Build healthy lactobacillus**
- **Improve wound healing**
- **Eliminate malnutrition**
- **Reduce radiation sickness**

Researchers are discovering how and why this algae is effective for human and animal health. Hundreds of scientific studies reveal how spirulina and its phytonutrients and extracts boost the immune system and improve health.

(*This information is solely for education and information purposes. It is not intended as medical advice. People with medical questions should consult their physician or health professional.*)

Summary of health benefits

Over the past 45 years, hundreds of published scientific studies have revealed the medical and therapeutic benefits of spirulina. Combined with the experience of millions of consumers, we know of a wide range of therapeutic effects from very low amounts of spirulina in the diet. We have indications that many new potential applications are yet to come. Here is a summary of these discoveries, followed by more in-depth discussion.

• **Anti-aging, brain health and neuroprotective effects.** Reduces inflammation, oxidative stress and improves brain function.

• **Anti-viral activity against HIV and Coronavirus.** Stronger resistance to all types of virus assaults. Reduces severity of infections and opportunistic infections.

• **Anti-cancer studies for prevention and inhibition.** Reduces risks of cancer and cancer tumors.

• **Anti-oxidant protection and anti-inflammatory effects.** Reduces oxidative stress, cellular damage and inflammation.

• **Beta carotene for healthy eyes and cancer prevention.** Corrects eye disease, lowers risks of cancer, reduces cancer tumors.

• **Cardiovascular heart health and cholesterol reduction.** Reduces hyperlipidemia, cholesterol, triglyceride and LDL levels.

• **Diabetes and hypertension reduction.** Lowers serum glucose levels and high blood pressure.

• **Gamma linolenic acid and prostaglandin stimulation.** Dietary GLA may help heart disease, PMS, obesity, arthritis and alcoholism.

• **Iron bioavailability and prevention of anemia.** More absorbable than typical iron supplements.

• **Athletic endurance and immunity.** Recovery from training stress and immune deficit.

• **Kidney and liver detoxification.** Helps the body eliminate toxicity from from heavy metals, drugs, chemical pollutants.

• **Lactobacillus improvement for better digestion.** Healthy flora means stronger immune system, reducing infections.

• **Malnutrition recovery.** Digestible protein, beta carotene and minerals for restoring healthy intestinal flora.

• **Phycocyanin and immune improvement.** Anti-viral, anti-inflammatory, builds blood cells, detoxifies organs.

• **Polysaccharides and immune improvement.** Anti-viral, raises antibody production and infection fighting T-cells.

• **Radiation sickness reduction.** Promotes evacuation of radionucleides from the body, reducing radiation sickness symptoms.

• **Sulfolipid anti-viral activity.** Blue-green algae sulfolipid extracts can stop HIV virus from attaching to cells, preventing infection.

• **Weight loss benefits.** Lowers appetite and reduces body weight.

• **Wound healing and skin health.** Applied topically on skin accelerates healing and skin health.

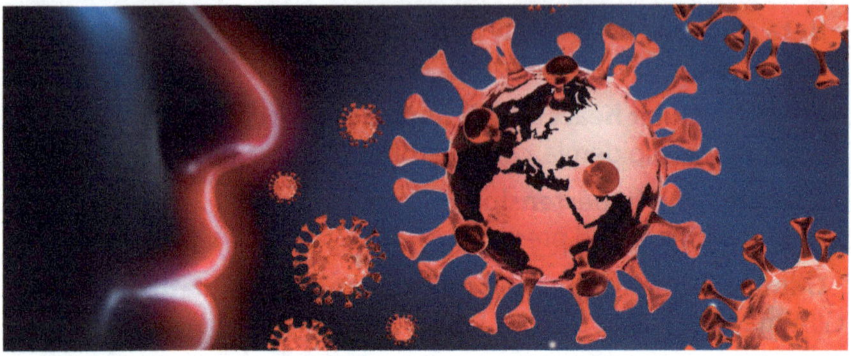

Spirulina and Covid-19, HIV, and Immune System

The first line of defense from viral infection - before masks, protective gear, social distancing and vaccines - is our own immune system. Vulnerable populations like the elderly, poor, disadvantaged and undernourished, and immune system compromised individuals, are at risk of chronic injury or death from Covid-19, HIV, flu, and the next emerging novel coronavirus.

Numerous scientific studies reveal how spirulina and its extracts strengthen the immune system and inhibit viral replication. Important parts of the immune system, bone marrow stem cells, macrophages, T-cells and natural killer cells exhibit enhanced activity.[1,2,3] Spleen and thymus glands show enhanced function. Spirulina causes macrophages to increase in number, become activated and more effective at killing germs.

Implications are clear: adding a small amount of spirulina to our daily diet will provide stronger resistance to all types of virus assaults and likely reduce severity of infections.

The following health benefits show spirulina and its components- phycocyanin, carotenoids, polysaccharides and sulfolipids- strengthen the immune system, reduce inflamation, inhibit viral replication, build cellular protection and accelerate detoxification of accumulated toxic and cancer causing substances.

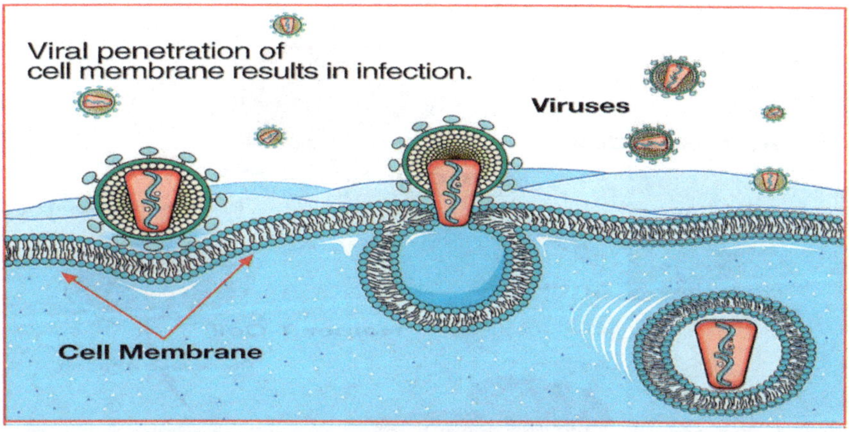

Spirulina prevents viral penetration of the cell membrane.

Potent Anti-Viral Activity

In 1996, Laboratory of Viral Pathogenesis, Dana-Farber Cancer Institute, Harvard Medical School and Earthrise Farms announced research, saying *"Water extract of Spirulina platensis inhibits HIV-1 replication in human derived T-cell lines and in human peripheral blood mononuclear cells. A concentration of 5-10 µg/ml was found to reduce viral production."*[1]

HIV-1 is the AIDS virus. Small amounts of spirulina extract reduced viral replication while higher concentrations stopped reproduction. Spirulina extract was non-toxic to human cells at concentrations stopping viral replication.

Scientists in Japan published studies about a water extract unique to spirulina named *Calcium-Spirulan.* It inhibits replication of HIV-1, Herpes Simplex, Human Cytomegalovirus, Influenza A virus, Mumps and Measles virus in-vitro yet is safe for human cells. It protects human and monkey cells from viral infection in cell culture. This extract *"holds great promise for treatment of HIV-1, HSV-1, and HCM infections, which is particularly advantageous for AIDS patients who are prone to these life-threatening infections."*[2]

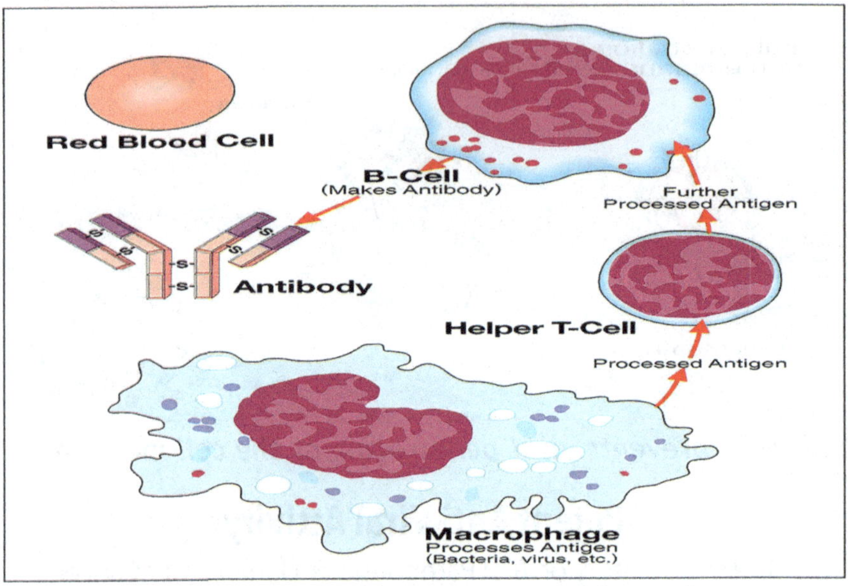

Key players in immunity stimulated by spirulina or its extracts.

Calcium-Spirulan is a polymerized sugar molecule containing both sulfur and calcium. Hamsters treated with this extract had better recovery rates when infected with a lethal Herpes virus.[3] The virus cannot penetrate the cell membrane to infect the cell and unable to replicate. It is eliminated by the body's natural defenses. Spirulina extracts could help patients lead longer lives.

Feeding studies show small amounts build up humoral and cellular arms of the immune system. Spirulina accelerates production of antibodies and cytokines, allowing the body to better protect against invading germs. The immune system includes T-cells, Macrophages, B-cells and anti-cancer Natural Killer cells that circulate in the blood and are rich in organs like the liver, spleen, thymus, lymph nodes, adenoids, tonsils and bone marrow. Spirulina up-regulates these cells and organs, improving the ability to function in spite of stresses from environmental toxins and infections.[4,5,6]

Spirulina benefits for HIV patients

About 34 million people worldwide are living with HIV, and 23.5 million live in Sub-Saharan Africa. The life expectancy of people living with HIV has improved due to highly-active anti-retroviral therapy py (HAART). However, only 36% of people in need of treatment in Sub-Saharan Africa have access to it.

Dietary supplements are often used to improve the nutrition of people living with HIV/AIDS. Research has shown spirulina benefits HIV-infected adult women. A 2014 study in Cameroon describes the effects of 5 g/day of Spirulina on 73 HIV-infected, adult females comparing spirulina with a placebo of equal protein content and energy.

Spirulina seemed to reduce the incidence of opportunistic infections and showed a positive effect on weight stabilization. This represents a significant benefit, since people with compromised immune systems such as HIV are vulnerable to opportunistic or concomitant infections.[1]

Spirulina extract may prevent serious Covid-19

Phycocyanin-rich spirulina extract was effective in inhibiting viral replications like HIV-1 in healthy volunteers. The anti-viral immunomodulatory and anti-inflammatory properties are attributable to the phycobiliprotein-phycocyanin complex. Available preclinical data supports the application of phycocyanin extracts as disease modifying therapeutics in Covid-19.

A 2020 Israeli study found an extract of spirulina may help Covid-19 patients avoid getting seriously ill. Spirulina reduces by 70% the release of an immune-system protein that causes dangerous cytokine storm in the lungs, leading to acute respiratory distress and organ damage. Clinical trials are planned next, with the goal of formulating oral spirulina drops.[1]

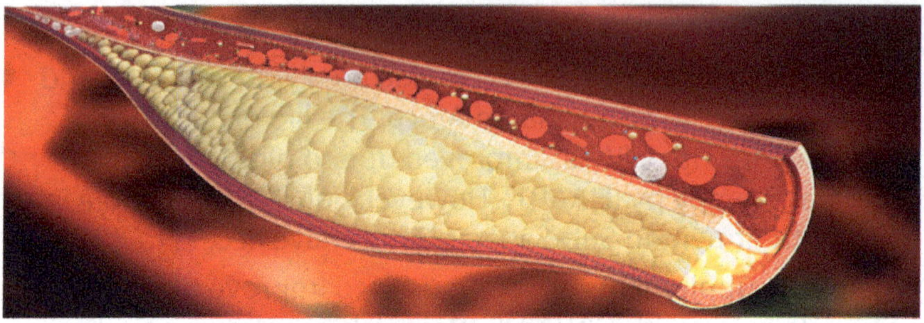
Cholesterol buildup in an artery.

Heart health and cholesterol reduction

People are aware of the need to lower cholesterol levels to lower risks of heart attacks and strokes. Besides dietary improvements, the search is underway to identify natural foods having a cholesterol reducing effect, such as fish oil or oat bran.

Spirulina is one of these foods. In Japan, 30 male employees with high cholesterol, mild hypertension, and hyperlipidemia showed lower serum cholesterol, triglyceride and LDL (undesirable fat) levels after eating spirulina for eight weeks. These men did not change their diet, except adding spirulina. Reduction of cholesterol was even greater in those men with the highest levels.

This study conducted by Tokai University concluded spirulina did lower serum cholesterol and was likely to have a favorable effect on alleviating heart disease since the arteriosclerosis index improved. No adverse effects were noted.[1]

West German research discovered cholesterol reduction during a weight loss study with spirulina.[2] Japanese research showed lower cholesterol without weight loss, suggesting that cholesterol reduction was not related to weight loss.[3,4]

A study in Mexico with 36 volunteers on serum lipids, glucose, blood pressure showed spirulina had a hypolipemic effect and reduced blood pressure.[5]

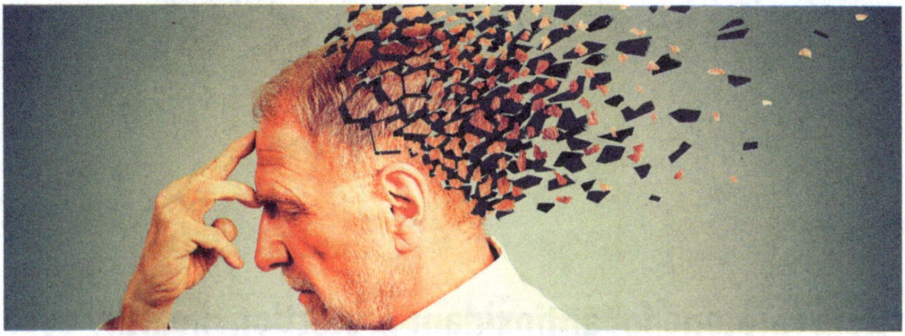

Anti-aging and neuroprotective effects

Some differences in the quality of life in old age and the onset of disease and aging depend on genes. However, diet and exercise play a role in how the central nervous system functions with aging. With aging comes normal age-related changes in the nervous system, including oxidative stress, inflammation, changes in learning, memory and neurotransmitter receptor function.[1]

Studies show diets with fruits and vegetables including spirulina support the aging central nervous system. Spirulina has high phycocyanin, chlorophyll and carotenoid pigments. Fruits and vegetables with deeper pigments supply more antioxidants. Blueberries have a high ORAC (Oxygen Radical Absorbance Capacity) value of 2,600, but spirulina is higher at 13,000. Studies with aged rats showed spirulina diets reduced inflammation and oxidative stress and improved receptor function in rat brains.[2]

Emerging evidence shows interaction between the gut microbiota, the immune system, and inflammatory pathways that influence aging. Reducing inflammation in the liver and intestines through gut microbiota modulation, may be key for healthy aging.[3]

Spirulina should be considered therapy for the aging brain. It has many actions to counterract oxidative stress and inflammation that occur as a consequence of aging and aid regeneration of the brain following injury or neurodegenerative disease."[4]

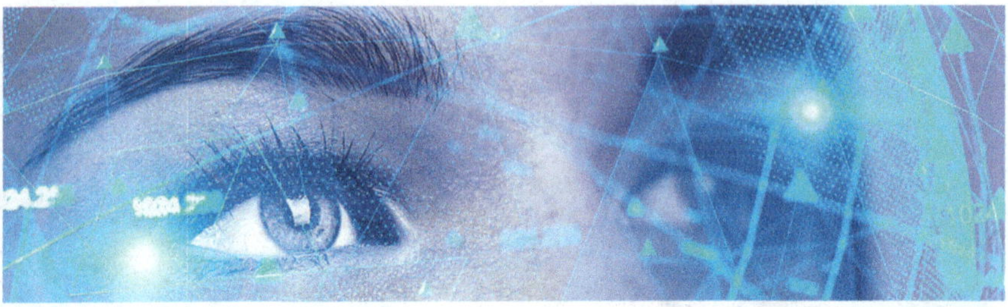

Beta carotene for antioxidant protection, healthy eyes

Beta carotene is the main source of Vitamin A for humans. Our bodies convert beta carotene to Vitamin A as we need it. Spirulina is the richest beta carotene food, over ten times more beta carotene than any other food, including carrots.

Beta carotene is one of the most effective substances for deactivating free radicals, which damage cells, leading to cancer. Free radicals are molecular fragments from environmental pollution, toxic chemicals, drugs, physical and emotional stress. Beta carotene prevents free radicals from reacting, and decreases risk of lung cancer, prevents chemically induced tumors in animals, prevents precancerous damage and builds immunological resistance.

In 1992, in the world's largest spirulina nutrition program with 5,000 children near Madras India consumed one gram a day for 150 days, providing the daily requirement of beta carotene (Vitamin A). *Bitot's spot,* scarring of the conjunctiva of the eye, decreased from 80% to 10%.[6] Spirulina was given to children in noodles, sweetened with sugar to preserve beta carotene.

Over 100 animal studies confirm Vitamin A and beta carotene inhibit development of cancers and tumors. Human studies correlate Vitamin A intake with decreased cancer risks.[2] Beta carotene (and not preformed Vitamin A from animal sources) correlated with lower cancer rates.[3] For those who do not eat 4-9 servings of fruits and vegetables daily, spirulina will add beta carotene insurance.

Reducing risks of cancers

Hundreds of studies have correlated lower incidence of lung cancer with beta carotene and Vitamin A. One study found the lower the serum level of beta carotene, the higher incidence of lung cancer.[1] Nine studies from 1974-1986 correlated lower digestive tract cancer (oral, stomach, colon, gastrointestinal) with beta carotene and Vitamin A. Two studies with women correlated lower breast and cervix cancer. A five year study in China with 29,000 people revealed daily doses of beta carotene, vitamin E and selenium reduced the incidence of cancer deaths by 13%.[2]

In 1982, *Diet, Nutrition and Cancer* published by the US National Research Council concluded "epidemiological evidence is sufficient to suggest that foods rich in carotenes or Vitamin A are associated with a reduced risk of cancer." It recommended a diet with beta carotene rich vegetables to reduce cancer risks.[3]

A 1987 Israeli study showed natural beta carotene is more effective than synthetic. Natural beta carotene is better assimilated by the body because it contains the 9-cis carotenoid isomer, lacking in synthetic carotene. This means beta carotene in algae and vegetables has greater antioxidant power than synthetic beta carotene.[4]

Anti-cancer tumor effects

Harvard School of Dental Medicine reduced oral cancer cells with spirulina extracts. A beta carotene solution applied to cancerous tumors in mouths of hamsters reduced tumors or caused them to disappear.[1,2]

Spirulina reversed oral cancer in pan tobacco chewers in Kerala, India. Regression of oral leukoplakia was found in 45% of those using one gram a day for one year, compared to 7% with placebo. Within one year of discontinuing spirulina, 45% of lesions returned. This was the first human study of its chemopreventive potential.[3]

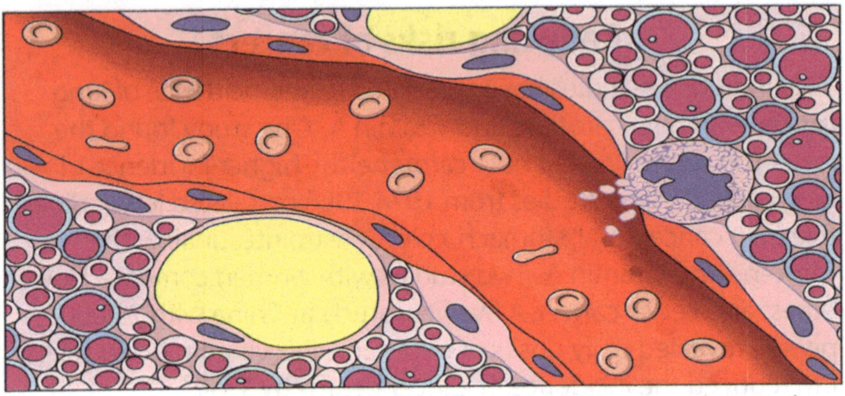

Spirulina or its extracts have shown effect in stimulating new red and white blood cells.

Phycocyanin builds blood

Spirulina has a dark blue-green color, because it is rich in brilliant blue phycocyanin. Studies show it affects the stem cells found in bone marrow. Stem cells are "Grandmother" to both the white blood cells that make up the cellular immune system and red blood cells that oxygenate the body.

Chinese scientists document phycocyanin stimulating hematopoiesis - creation of blood, emulating the hormone erythropoietin. EPO is produced by healthy kidneys and regulates bone marrow stem cell production of red blood cells. Phycocyanin regulates production of white blood cells, even when bone marrow stem cells are damaged by chemicals or radiation.[1]

Based on this effect, spirulina is approved in Russia as a "medicine food" for treating radiation sickness. The Children of Chernobyl suffer poisoning from food grown on radioactive soil. Their bone marrow is damaged, making them immunodeficient. Radiation damaged bone marrow cannot produce normal red or white blood cells. The children are anemic and suffer from allergies. Children fed five grams in tablets each day made recoveries within six weeks.[2]

Phycocyanin enhances the immune system

Part of the global effort to identify natural substances with an immune system boosting or anti-cancer effect focuses on spirulina's blue protein pigment, phycocyanin.

In Japan, phycocyanin was taken orally by mice with liver cancer. Survival rate of the phycocyanin group was higher than the control group. After five weeks, 90% of the phycocyanin group survived, but only 25% of the control. After eight weeks, 25% of the phycocyanin group still survived, yet none of the control was alive. This suggests phycocyanin may increase survival from cancer.

In another study, after two weeks white blood cells (lymphocyte activity) of a phycocyanin group were higher than the control and higher than or equal to a normal group without cancer. This suggests phycocyanin raises lymphocyte activity.[1]

The lymph system's function is to maintain healthy organs in the body, and protect against cancer, ulcers and other diseases. Phycocyanin acts by strengthening the body's resistance through the lymph system. Phycocyanin may be active in preventing degenerative organ diseases by boosting immunity.

A Japanese patent states a small dosage of phycocyanin daily accelerates normal control cell functions that prevents generation of malignancy such as cancer or inhibits its growth or recurrence.[2] The patent recommends a phycocyanin dosage of 0.25 to 2.5 grams per day. Spirulina with 15% phycocyanin would mean consuming 1.7 to 17 grams per day, or about 10 grams.

Polysaccharides enhance the immune system

In 1979, Russian scientists published initial research on the immune stimulating effects on rabbits from lipopolysaccharides in spirulina.[1] More recent studies in China and Japan have shown polysaccharide extracts increased macrophage function, antibody production and infection fighting T-cells.

In 1991-94 in China, spirulina polysaccharides and phycocyanin increased immunity in mice by enhancing bone marrow reproduction, growth of thymus and spleen and biosynthesis of serum protein.[2,3,4,] Hamsters treated with a polysaccharide extract had better recovery rates against a herpes virus. A water extract unique to spirulina, *Calcium Spirulan*, inhibited replication of HIV-1, Herpes Simplex and other viruses, yet was safe for human cells.[5]

In the USA, a water soluble extract increased macrophage activity in chickens. In further studies, chickens fed a diet with less than 1% spirulina showed improved immune performance without any adverse side effects. The whole immune system array of killer cells, helper cells and antibody production was supercharged.[6,7] Similar benefits were found for cats.[8]

Researchers are testing spirulina and its extracts for a broad spectrum vaccine against bacteria. Because it is a safe natural food, this research has created a sensation among animal scientists. They are scrambling to replace ineffective antibiotics with probiotics that strengthen the immune system and prevent disease. Based on animal research, 3 grams per day may be effective for humans.[9]

In 1996, U.S. scientists announced research showing a water extract of spirulina inhibits HIV-1 replication in human derived T-cells.[10] HIV-1 is the AIDS virus. Small amounts of the extract reduced viral replication, while higher concentration stopped reproduction. The extract seemed to prevent the virus from penetrating the cell membrane and the virus was unable to replicate.

Sulfolipid extracts stop the HIV virus

National Cancer Institute (NCI) is searching the world for natural plants and organisms that have biologically active anti-cancer agents. The famous periwinkle plant in the Madagascar rain forests is one example of a new cure. Having scoured terrestrial organisms, scientists are now looking towards the sea.

NCI scientists have screened 18,000 extracts of marine organisms for activity against tumors, viruses and fungi and for immune system stimulation properties. Extracts of sea squirts, sea whip soft corals, and sea sponges offer potential drugs.

In 1989, the NCI announced that chemicals from blue-green algae were found to be "remarkably active" against the AIDS virus.[1] These are the naturally occurring sulfolipid portions of the glycolipids. Sulfolipids can prevent viruses from attaching to or penetrating into cells, thus preventing viral infection.

NCI emphasized that a larger testing program including tests on humans with the AIDS virus would not begin until sulfolipids can be obtained in much larger quantities. These scientists further speculated that if sulfolipids proved effective, used in combination with drugs like AZT, they would be safer and more effective.

Scientists used extracts of the blue-green algae lyngbya, phormidium, oscillatoria, and anabaena. Spirulina is known to contain glycolipids and sulfolipids.[2] It contains 5-8% lipids, and of that, about 40% are glycolipids, and 2-5% are sulfolipids.[3] Blue-green algae can be cultivated to increase the lipids, and presumably, the sulfolipids. This means it could be grown on a large scale for extraction of this valuable anti-cancer and anti-AIDS substance.

In 1996, NCI scientists announced another extract from the blue-green algae nostoc, *cyanovirin-n*, could be a broad spectrum virucidal agent against HIV. This unique antiviral protein was selected for further high-priority development.[4]

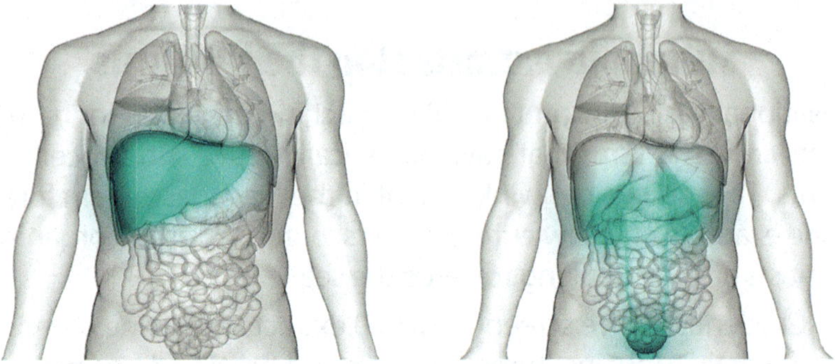

Kidney and liver detoxification from heavy metals, drugs, chemical pollutants

Kidneys play an essential role in cleansing the body of toxins. Heavy metals and many drugs are known to be toxic. Scientists are interested in substances that cleanse the kidneys of side effects from heavy metal poisoning or from medicines or pharmaceuticals.

In Japan, spirulina reduced kidney toxicity from mercury and three pharma drugs in laboratory rats.[1] Scientists measured kidney toxicity- blood urea nitrogen (BUN) and serum creatinine. When rats were fed a diet with 30% spirulina, BUN and serum creatinine levels decreased dramatically. In a follow-up study, the compound suppressing renal toxicity was the extract phycocyanin.[2]

These studies suggest spirulina may be beneficial for humans suffering from heavy metal poisoning. Kidney side effects from pharma drugs may decrease when it is eaten along with the administration of drugs. Side effects limit the dosage of many drugs, slowing recovery. Higher dosage of drugs and shorter recovery times may be possible. Study of the kidney cleansing effect offers an insight into cleansing people have reported while fasting.

Another study found liver detoxification of chemical dioxins. The fecal excretion of dioxin was 7 to 11 times higher for rats on a diet of chlorella, spirulina and chlorophyllin and suggested a useful new approach in treatments.[3]

Builds healthy lactobacillus

Healthy lactobacillus in the intestines provides humans with three major benefits: better digestion and absorption, protection from infection, and stimulation of the immune system.

In Japan, research showed spirulina increased lactobacillus in rats 3 times over a control group. A diet with 5% spirulina for 100 days showed 1) caecum weight up 13%, 2) lactobacillus up 327%, and 3) Vitamin B1 inside the caecum up 43%.[1] Since spirulina did not supply additional B1, it improved B1 absorption. It should increase lactobacillus in humans and raised absorption of vitamins.

This has implications for AIDS. Some researchers believe the inability to absorb nutrients in the intestines can cause serious immune deficiency. The absence of lactobacillus leads to thriving infections. Malabsorption with opportunistic infections can initiate full-blown AIDS. One strategy for halting progression of AIDS is using supplements (to correct malabsorption) and lactobacillus (to maintain intestinal flora and prevent infection).[2]

Wound healing and antibiotic effects

People have used spirulina in face creams and body wraps, and in baths to promote skin health. The Kanembu in Chad use harvested algae as a skin poultice for treating diseases.[1,2] Pharma compounds in France containing spirulina accelerated wound healing. Patients used spirulina and extracts in creams, ointments, solutions and suspensions.[3] In Japan, cosmetic packs with spirulina and its hydrolyzates promoted skin metabolism and reduced scars.[4]

Effects against diabetes and hypertension

Spirulina may have a positive effect against diabetes. A water soluble fraction was found effective in lowering serum glucose level at fasting while the water-insoluble fraction suppressed glucose level at glucose loading.[1] It may reduce blood pressure. In a recent study with rats, it was found to reduce high blood pressure.[2]

Iron bioavailability and correction of anemia

Iron is the most common mineral deficiency worldwide, but typical iron supplements are not well absorbed by the body. Spirulina fed rats absorbed 60% more iron than rats fed a conventional iron supplement, suggesting a highly available form of iron in spirulina.[1] An earlier study showed it corrected anemia.[2]

In Japan, 8 young women had been limiting meals to stay thin, and showed hypochronic anemia – lower than normal blood hemoglobin content. After 4 grams of spirulina after each meal, in 30 days, blood hemoglobin content increased 21% to a satisfactory level, no longer considered anemic.[3]

Athletic endurance and immune deficit

Athletes in intensive training can suffer from non-anemic iron deficit, with exhaustion and muscle fatigue. A 1998 study with Macedonian athletes taking spirulina for two months showed a rise in iron reserves. This dietary modification can eliminate iron deficit symptoms and optimize athletic health and physical capacity.[1]

A 2018 study was conducted with 19 members of the Polish Rowing Team. The results imply that supplementation with spirulina extract may protect athletes against a deficit in immune function (especially, anti-infectious function) associated with strenuous exercise, and may cause a beneficial shift in "overtraining threshold" preventing a radical deterioration of immunity.[2]

Margarite at Maison de la Nutrition, Antsirabe, Madagascar. (Antenna-France.)

Benefits for malnourished children

As little as 10 grams a day brings recovery from malnutrition, especially for infants. Spirulina was given to undernourished children in Mexico[1] and adults[2] with beneficial results. It was more than 10% of their diet with no adverse effects.

In Togo, recovery of malnourished infants was reported in a village clinic. Children given 10 to 15 grams per day mixed with millet, water and spices, recovered in several weeks.[3] In India, studies with preschool children showed spirulina carotenes helped children recover from Vitamin A deficiency.[4,5] In Romania, tablets were given to patients in a Bucharest clinic, suffering from weight loss in conjunction with pancreatitis, rheumatoid arthritis, anemia and diabetes. The patients gained weight and health improved.[6]

In China's Nanjing Children's Hospital, spirulina was mixed in a 'baby nourishing formula' with baked barley sprouts. 27 of 30 children aged two to six recovered in a short period from bad appetite, night sweat, diarrhea and constipation. Researchers concluded this is a genuine health food for children.[7] In another study, children deficient in the essential mineral zinc, made more rapid recovery with high zinc spirulina than a standard zinc supplement.[8]

Reduces radiation effects for Children of Chernobyl

Years after the Chernobyl disaster, the water, soil and food over an 11,000 square mile area were still contaminated. Over 160,000 children have been victims of radiation poisoning, with birth defects, leukemia, cancer, thyroid disease, anemia, loss of appetite and depressed immune system, now called "Chernobyl AIDS."

Doctors reported spirulina's health benefits for child victims of Chernobyl radiation. Spirulina reduced urine radioactivity levels by 50% in 20 days. This result was achieved by giving 5 grams a day to children at the Minsk, Belarus Institute of Radiation Medicine. The Institute program treated 100 children every 20 days.[1]

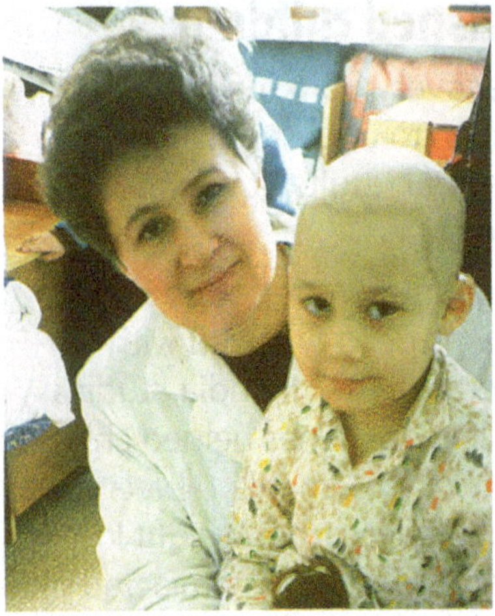

A 1993 report concluded "spirulina decreases radiation dose load received from food contaminated with radionucleides, Cesium-137 and Strontium-90. It is favorable for normalizing the adaptive potential of children's bodies in conditions of long-lived low dose radiation."[2]

Nurse and child radiation victim, at a medical clinic in Belarus.

In 1990, the Belarus Ministry of Health concluded spirulina promotes the evacuation of radionucleides from the human body with no side effects registered. The Ministry considered this food was advisable for the treatment.

In a 1991 study of 49 kindergarten children aged 3 to 7 years old in Beryozova, spirulina was given to 49 children for 45 days. Doctors found T-cell suppressors and beneficial hormones rose, and in 83% of the children, radioactivity of the urine decreased.[3]

Russian Patent for radiation sickness

A Russian patent was awarded in 1994 for spirulina as a medical food to reduce allergic reactions from radiation sickness, based on a study of 270 children living in radioactive areas. They had radiation sickness and elevated levels of Immunoglobulin (IgE), a marker for high allergy sensitivity. Thirty five were prescribed 20 tablets per day (5 grams) for 45 days. Consuming spirulina lowered the levels of IgE in the blood, which normalized allergic sensitivities.[1]

Research through1999 in Belarus showed immune building and detoxifying effects in children. Scientists theorized spirulina may form non-absorbable complexes of radionucleides through analogues such as calcium or potassium to promote excretion.[2]

GLA and prostaglandin stimulation

Foods high in saturated fats may block beneficial essential fatty acids, leading to disease. Gamma linolenic acid (GLA) is a precursor for the body's prostaglandins, master hormones that control many functions. The prostaglandin PGE1 helps regulate blood pressure, cholesterol synthesis, inflammation and cell proliferation.[1]

Studies show dietary intake of GLA can help arthritis,[2] heart disease,[3] obesity[4] and zinc deficiency.[5] Dietary GLA may help heart disease, PMS, obesity, arthritis and alcoholism.[6] In Spain, GLA in spirulina and evening primrose oil is prescribed for treatment.

Known sources of GLA include two foods, human milk and spirulina, and oil of the evening primrose plant, black currant and borage seeds. Ten grams of spirulina has over 100 mg of GLA.[7]

Weight loss research

Researchers in Germany used 15 human volunteers to test appetite reduction. Obese outpatients following a weight reduction diet took spirulina tablets. About 6 tablets three times a day over four weeks showed a small but significant weight reduction.[1]

Algae medicines from bioactive compounds

In *Emerald Renaissance*, Mark R. Edwards describes the vast potential of bioactive algae compounds.[1] Algae provide a superb biomanufacturing platform for a range of industries including bioenergy, biopharmaceuticals, biomaterials, nutraceuticals, agriculture, health, cosmetics and personal care. Biomanufacturing with algae allows low cost production, safety and scalability.

"Algae medicines, with and without genetic modifications, will transform both disease prevention and treatment. Medical bioactive compounds can be found and grown in algae faster, at lower cost and with higher quality than terrestrial plants."

"Individualized medicines will advance algae biofactories with compounds tailored to the DNA of each consumer. Development times for land plants take far too long for individualized medicine. Algae production platforms are able to produce bioactive compounds quickly and efficiently to support individualized medicine."

Scientific health library at SpirulinaSource.com

Scientific studies available online

Many scientific health abstracts and complete papers for download are available online at SpirulinaSource.com. The curated Health Library has over 100 references by health category covering 45 years of international research.

Chapter 5

The variety of products around the world

- **Market evolution in the USA**
- **Health food supplements around the world**
- **Healthy food ingredient for drinks, pasta, snacks**
- **Natural blue color for foods and cosmetics**
- **Specialty feed for fish, aquaculture and animals**
- **Pet food for birds, cats and dogs**
- **Fluorescent markers and ezymes for research**
- **Super spirulina products and extracts**

In over 40 countries, spirulina tablets, powder and capsules are widely used. Spirulina is an ingredient in pasta, cookies, snack bars and juice bars, in personal care products like skin creams and shampoos, and in pet supplements for fish, aquatic animals, birds, cats and dogs. Natural blue food color from spirulina is replacing chemical blue. More nutraceutical extracts are coming.

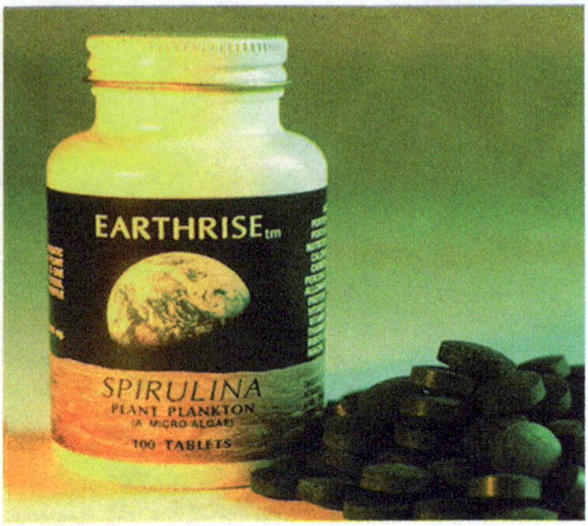

Article in San Francisco Examiner. First tablets.1979.

Market Evolution in the USA

The first microalgae sold as a health supplement in modern society was chlorella, beginning in the 1970s. Spirulina began selling in Japan as a food supplement in the late 1970s. In the USA, Earthrise® Spirulina was introduced in 1979 in natural food stores.

Introduction 1979-1981

About 10% of all U.S. consumers were eating some health foods. Concerned about the decline in food quality from over-processing, chemical additives, pesticides and pollutants, they opted for natural foods. These people embraced algae, discovering many ways to use it. Spirulina gathered a small, dedicated and growing following.

Diet boom 1981-1983

In 1981, a national tabloid headlined: *Doctor's Praise: A Safe Diet Pill – You'll Never Go Hungry.*[1] It claimed spirulina was a safe and effective appetite suppressant. Thousands of people lined up at stores to try this latest 'magic' diet pill. Spirulina became a new diet trend. News passed by word of mouth, magazines, newspapers, radio and TV. Diet pill companies jumped on the bandwagon. New brands appeared. But there wasn't enough spirulina to go around.

Earthrise® Spirulina ads: 1981, 1990, 1997.

Popularity subsides 1984-1986

World spirulina supply was about 500 tons per year from Mexico and Thailand. Much was sold in Japan, and only several hundred tons per year of real spirulina actually entered the US market. Because of the shortage, many new brands sold by diet pill companies were adulterated, having less than claimed on the label. Some contained none or were adulterated with green fillers. Hoping to magically lose weight, most people paid for green pills that were not the real thing, and experienced little health benefit.

Coming back 1987-1990

Natural food consumers who knew how to use algae properly and enjoy its benefits remained faithful. As more people recognized the real health value, demand rose again, along with green superfoods such as chlorella, barley grass and wheat grass. Green superfoods grew about 30% per year in the 1990s. New products appeared with herbs, phytonutrients, vitamins and minerals to raise energy, boost athletic performance and endurance, promote a lighter appetite and antioxidant protection. Companies created meal replacements with spirulina, snack bars, pasta, fruit and vege juices.

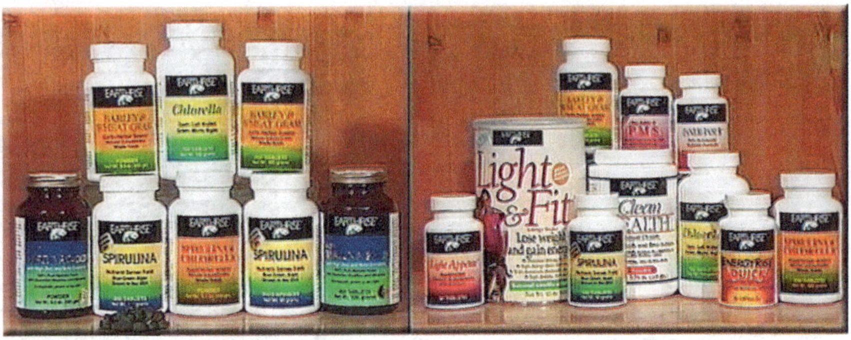

Earthrise® product line 1990s.

Decades of growth 1991-2020

People became aware that nutraceutical foods have benefits beyond nutrition, and foods with phytonutrients can help prevent disease. As more research on spirulina health benefits were published, it became known as a potent superfood.

With new US FDA dietary supplement label regulations in the 1990s, spirulina products could show label claims about health, structure and function benefits. New labeling and publicity propelled food and herb supplements into mainstream markets.

Spirulina today

Much spirulina is now produced in China and India for markets in Asia, Europe and North America. Spirulina is widely used the US market for health and dietary supplements and as an ingredient in human foods, animal and pet foods, and for blue food color.

Algae in Products Today

Algae is in thousands products today

Today, spirulina and other algae are ingredients in thousands of products for food, feed, colors, nutraceuticals, medicinals, cosmetics and personal care, biofertilizers and fine chemicals. Even more innovative algae based products are coming.

Algae Products and Markets

Food Supplements	with microalgae, marine algae, dha and epa oils
Health Food Ingredient	in protein shakes, juice drinks, energy bars
Personal Care Products	creams, masks, shampoos, cleansers. cosmetics
Nutraceuticals, Medicinals	pigments, oils, antioxidants, medical diagnostics
Edible Seaweeds	nori, dulse, wakame, kelp, seasoning
Food Ingredient	proteins, pigments, thickeners, stabilizers, oils
Pet Food Ingredient	aquarium fish and specialty pet foods
Fish Feed Supplement	for survival, health, nutrition, coloration, omega 3
Animal Feed and Oils	for survival, health, nutrition, omega 3
Fertilizers	plant food, growth promoters, soil conditioners
Biopolymers	packaging, bioplastics, adhesives
Fine chemicals	industrial enzymes, esters, resins

Spirulina around the world

Spirulina Viva crunchies, fresh, snacks. Mexico.

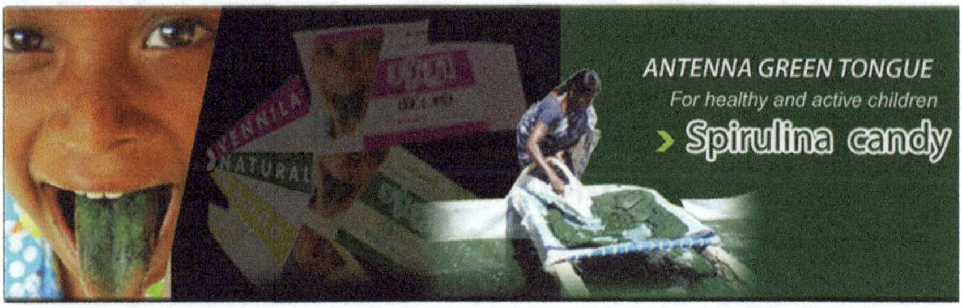

Antenna Green Tongue candy. India.

Advertisement for spirulina. Korea.

Simplicity spirulina products. India.

June spirulina products. Myanmar.

Plate with Oikos tablets. Spain.

Spirulina as a healthy food ingredient

Spirulina and green superfoods are popular in green drinks and protein bars and used in chips, dips, crackers, noodles, pasta and foods for color, nutrition or functional effect.

An ingredient for drinks, snacks and foods.

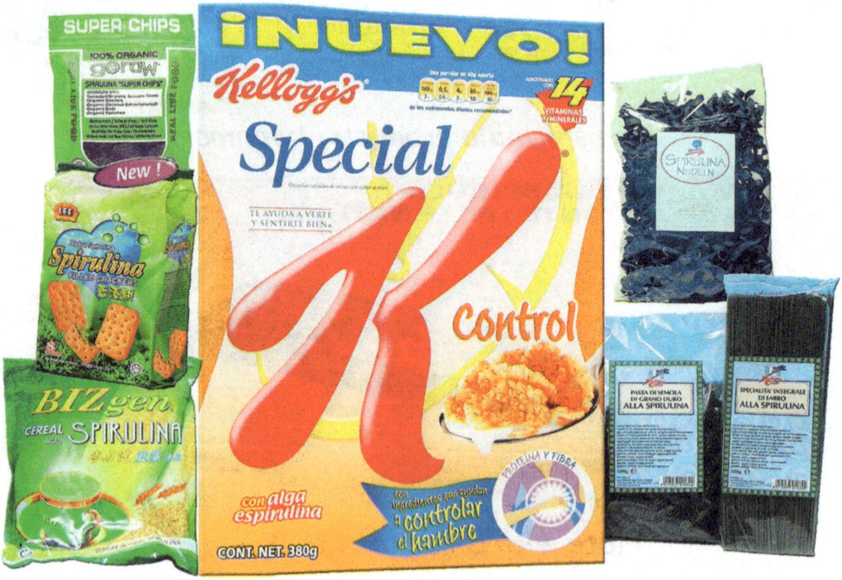

Pasta.Italy • Spices.France • Bread.Netherlands.

June beauty products and spirulina beer. Myanmar.

Energaia fresh spirulina and pasta. Thailand.

Phycocyanin products. Yuki enjoys gum (H. Shimamatsu). Nestles Smarties.

Spirulina blue color for foods, drinks, cosmetics

If you mix spirulina powder in water, within an hour a blue color appears. This water-soluble color is phycocyanin. Spirulina contains 10-20% phycocyanin by weight. In Japan, regulations mandate natural food colors. Dainippon Ink & Chemicals was the frst to extract blue food color from spirulina for chewing gum, ice sherberts, popsicles, candies, soft drinks, dairy products and wasabi, the green colored hot paste served in sushi bars.[2] Another form is made for natural cosmetics.

Drinks, juices and cocktails with spirulina blue color.

Blue food bowls, pancakes, ice cream, desserts.

Blue drinks and foods are trending

The U.S. FDA approved spirulina phycoyanin as a blue food coloring in 2013 and the EU approved it as well. By 2020, adding this safe natural blue color to drinks and foods had become popular.

New technology brings 'super spirulina'

Farms are growing spirulina with higher nutritional attributes. Because microalgae reproduce so rapidly, changes while growing can increase specific phytochemicals.

Higher phycocyanin, iron and trace minerals

By selecting the species and adapting growing conditions, some farms produce spirulina with high 15-20% phycocyanin content for color extraction. Algae absorbs minerals added to pond water while growing and some spirulina is grown for higher levels of iron, zinc and selenium. Zinc is essential for good health, powering antioxidant enzymes and a strong immune system. Beta carotene and zinc work synergistically to protect against free radical damage.

Phytochemical and pharmaceutical extracts

Scientists are searching for phytochemicals and pharmaceuticals in algae. Because studies have shown therapeutic effects, extracts of the active ingredients are promising. Growers will learn how to raise concentration of these substances.

Fish foods for koi and other colorfish contain spirulina.

Specialty food for fish, birds, animals and pets

Demand is surging for specialty aquaculture feeds that increase growth rates and disease resistance for farmed fish and prawns. Tropical fish, ornamental birds, animals and pets consume a portion global spirulina production.

Color beauty for champion koi

Highly prized fancy koi carp have distinctive bright red, yellow, orange, white and black markings, bred by hobbyists and fish culturists in Japan and China for centuries. If well taken care of, koi can live as long as people, becoming family pets for generations.

Champion fish can sell for tens of thousands of dollars each, so feeding the right food is a good investment. Many koi feeds include 5 to 20% spirulina for its rich carotene pigments that enhance red and yellow patterns, while leaving a brilliant pure white. This clarity and color definition increases their value.

Spirulina – Benefits for aquaculture

Species	Benefits
Aquariums	Provides the base link of the food chain for a thriving tank and zooplankton community.
Ayu fish	Improves flavor and texture of meat, improves color, reduces disease.
Baby fish	Increased resistance to parasites, viral and bacterial diseases due to enhanced cellular immune system function. Stimulates feeding, better survival, improved appearance.
Brine shrimp	Replaces live algae for growth. Use as enrichment before feeding to fish or prawns. Improves color.
Fancy koi carp	Enhances skin quality, color, shine and disease resistance. Improves appearance.
Fish & marine larvae	Increases appetite, nutrition, growth and survival rates.
Milkfish	Improves growth and reproduction.
Prawns	Replaces live algae. Increases resistance to disease, stimulates growth, improves reproduction and enhances color.
Queen conch	Improves growth in nursery, reduces need for live algae.
Salmon	Enhances disease resistance, improves appetite, skin quality, color, reproduction.
Tilapia	Protects from parasite infection and bacterial disease. Enhances cellular immune system function, coloration and appetite.
Tropical fish	Enhances skin quality, color, shine and disease resistance. Improves appearance.
Yellowtail tuna	Reduces disease, increases growth and survival. Enhances coloration.

Courtesy of Ronald Henson. PO Box 459, Tollhouse CA 93667 USA.

Spirulina products for fish and shrimp hatcheries.

Popular for aquaculture products

Fishing fleets are rapidly depleting ocean fish, threatening ocean food chains. To offset the dwindling wild catch, the fish farming industry is growing at 10% per year. Adding spirulina to fish feeds helps solve the two biggest problems. First, farmed fish are susceptible to disease. Second, the flavor, texture and skin color are often inferior to the wild fish.

Growers of yellowtail tuna, a popular sushi fish, know it increases survival and growth rates of the hatchery fry and improves yellow side lines, skin color and growth rate. Prawn growers feed prawns spirulina just before harvesting to enhance their splendid colors for appeal in sushi bars.

Japanese fish farmers discovered five key benefits to using feeds with spirulina: 1) better growth rates, 2) improved quality and coloration, 3) better survival rates, 4) reduced medication, 5) reduced waste in the effluent.[3]

Growth rates increase and less feed is wasted because spirulina increases palatability of the entire feed. Fish respond to its flavor. They grow faster, taste better and resist disease. Environmental regulations make it vital to reduce effluent pollutants. Spirulina helps fish farms improve effluent without costly treatment systems.

Many aquarium fish foods contain spirulina.

Improves survival of fish fry and brine shrimp

Aqiaculture fish are grown from tiny fry. This first stage is difficult and critical for success. Often survival rates are low. Spirulina added to the feed at 1 to 10% levels increases survival rates, allowing fish to reach market sooner. It is the best food for tiny brine shrimp, a popular food for aquarium fish. Tiny zooplankton are another delicious food for larger fish, but growers have found them hard to cultivate. When fed spirulina, chances improve greatly.

Health food for the aquarium

Spirulina promises five benefits for healthy aquarium fish: 1) concentrated vitamins and minerals, 2) rich in muco-proteins for healthy skin, 3) phycocyanin for reduced obesity and better health, 4) essential fatty acids for proper organ development, 5) rich coloring agents such as carotenoids. Feeding will result in beautiful, healthy and longer-lived fish.[4]

Spirulina will not grow in the aquarium. Hobbyists can find floating flake food with spirulina in pet shops. Some use frozen food or make their own fish foods. Public aquariums use spirulina to feed marine fish, invertebrates and raise daphnia and brine shrimp to feed to their fish.

Spirulina supplements for birds.

Health, beauty and color for ornamental birds

Zoos feed flamingos and ibis spirulina for feather color and shine, healthy beaks and skin, and good bacteria in the digestive tract. Birds are healthier, without drugs or chemicals.[5] Pet birds are more beautiful, healthier and live longer. Spirulina is concentrated; use at 1 or 2% of diet. Ailing birds may be fed more, not exceeding 5%. Sprinkle powder over soft food or mix into hand feeding formula.

Enhanced fertility and health for bird breeders

Canary, finch, parrot, and lovebird breeders use spirulina to increase coloration, accelerate growth, sexual maturity and raise fertility rates. Ostrich and turkey breeders use it to increase fertility and reproduction. It enhances desirable yellow skin coloration in chickens and the deep yellow color of egg yolks. Adding a percentage to chicken diet stimulates macrophage production, improving immune performance and disease resistance without side effects.[6]

Tonic for horses, cows and breeding bulls

Racehorse owners use spirulina in their feed for faster times and recovery. Research has shown spirulina fed to horses reduced oxidative stress and inflammation and may help treat equine metabolic syndrome (EMS).[7] Dairy farmers keep cows healthy by improving intestinal flora. It increases sperm count of breeding bulls and fertility in females, raising the reproduction rates.

Spirulina—Benefits for pets and animals

Pets	Benefits
Bird	Improves feather quality, color, fertility, speeds development of immune system in babies for better resistance to infection.
Cat	Improves coat, healthy skin, healing, resistance to viral infections and cancer.
Dog	Improves coat, healthy skin, reduces dermatitis, improves disease resistance.
Horse	Reduces anemia, builds red and white blood cells, improves appetite, energy, resistance to infection, recovery from injury. Shiny coat.
Parrot	Improves feather quality, color, fertility, speeds development of immune system in babies for better resistance to infection.
Rabbit	Improves resistance to respiratory infection.
Reptile	Improves skin gloss, cleaner shedding, disease resistance.
Animals	*Benefits*
Dairy Cattle	Short-term therapeutic for mastitis, stimulates appetite.
Poultry	Reduces early poultry mortality.
Swine	Improves disease resistance in weaner pigs.

Courtesy of Ronald Henson. PO Box 459, Tollhouse CA 93667 USA.

Healthy food for cats and dogs

For healthy skin and lustrous coat, spirulina is good for cats and dogs, for nursing mothers, bottle-fed kittens and puppies. Appetites of finicky cats perk up with a little sprinkled on their food. Spirulina fed pets have a fresher breath odor.

Use a spoonful of powder and sprinkle over soft moist food. A little goes a long way. Use at about 1% of daily food intake. A pinch each day for small cats and kittens; for large cats and nursing mothers use 1/4 tsp. For small dogs use 1/4 tsp each day; medium dogs use 1/2; large dogs use 1. Use 1/2 tablet a day for small cats and kittens; large cats and nursing mothers use 1 tablet. For small dogs use 1/2 or 1 tablet; medium dogs use 2; large dogs 3-4.

Fluorescent markers for medical tests

Purified algae extracts are used to track diseases in the human body. Phycobiliproteins are natural pigments in red and blue-green algae, including spirulina. When purified, they are fluorescent, stable, and water soluble. Fluorescent dyes improve detection of cancer, screen blood for AIDS and monitor blood levels of drugs.

Researchers watch fluorescent markers in a microscope and follow their path inside the human body. Linked to molecules such as monoclonal antibodies, when these antibodies attach to receptor sites on cells or tissues, cells can be viewed by the glowing dye marker illuminated by light.[8] Phycobiliproteins are used in medical diagnostic kits and are 10-30 times more intense than dyes and are alternatives to radioactive markers.[9]

Enzymes for genetic research

Living spirulina contains many enzymes and three restriction enzymes have been discovered. Genetic engineering researchers use restriction enzymes to cut DNA at precise locations[10]. One restriction enzyme in spirulina called *Spl-1* is not found in any other microbe, bacteria, fungi or algae. Japanese scientists extract *Spl-1* from living spirulina and sell it as a reagent for genetic research.

One theory proposed to explain spirulina's long life is the role of restriction enzymes. By cutting the DNA of invading enemy microbes, spirulina may protect itself from bacteria or viruses.[11]

Polysaccharide extracts for fish and animals

Several aquaculture feeds offer polysaccharides and beta glucans from yeast to enhance immune activity to replace antibiotics used to control disease in farmed animals. Two problems with antibiotics are antibiotic resistant microorganisms and accumulation of antibiotics in fish. Researchers are studying the immune properties of algae polysaccharide extracts for therapeutic uses.

USA regulatory status and issues

Food: American grown spirulina produced by Earthrise and Cyanotech has GRAS Status (Generally Recognized as Safe). This was based on the published information on the safety of spirulina and a description of their GMP and Quality Assurance Program and US Food and Drug Administration (FDA) review. Spirulina is regulated as a food and dietary supplement. In 1994, the U.S. Congress passed the Dietary Supplement Health and Education Act (DSHEA). Dietary supplements, like spirulina, may make health statements about the structure and function of the body on product labels based on evidence.

Feed: As an animal feed supplement, spirulina does not yet have an AAFCO (Association of American Feed Control Officials) ingredient description, meaning it should be used in feed applications which are low priority enforcement. Since spirulina is already GRAS, AAFCO status should be achievable.

Food Colors and Additives: The US FDA has approved phycocyanin extract of spirulina for use as blue food colorant. FDA regulates food colors and additives and requires a long approval process before natural colors can be used in foods and cosmetics.

Pharmaceutical Extracts:

Biotechnology and pharmaceutical drug companies have become interested in algae for extracts to be used in nutraceutical and pharmaceutical drugs. Algae extracts must pass rigorous screening for toxicity and lengthy regulatory review before approval is granted for use.

Chapter 6

How spirulina is ecologically grown

- **Tour a large commercial farm**
- **Worldwide spirulina farms**
- **Photobioreactors and tubular systems**
- **Rooftop and urban farms**
- **Emerging spirulina microfarms**

By growing spirulina in designed ponds under controlled conditions, a pure culture can be maintained, which is not possible in natural lakes.

Large commercial farms have evolved in the United States, Thailand, Taiwan, Japan, Mexico, China, India and other countries. Although new technologies are coming, almost all large spirulina farms today use open raceway pond systems with paddlewheels introduced over 50 years ago.

Small microfarms are now proliferating around the world, selling innovative products in their local region.

Production ponds with food grade liners. Each is 5000 square meters-larger than a football field.

Quick tour of Earthrise Nutritionals[1]

In 1977, Proteus Corporation, the first U.S. algae entrepreneurs began pilot ponds in California's Imperial Valley and in 1979 introduced Earthrise® Spirulina. In 1981, Proteus formed a partnership with DIC, a Japanese corporation, and founded Earthrise Farms, sharing a common vision of microalgae's impact on the world.

One of the world's largest spirulina farms

Earthrise becamethe world's largest spirulina farm, expanding to cover 108 acres, supplying 40 countries with spirulina. With 37 growing ponds mixed with 50 foot paddlewheels Earthrise produces 500 metric tons of dry powder per year.

Ecological pond cultivation

Ecological farming produces healthy and unpolluted foods. In this sunny part of California, mineral rich Colorado river water, which supplies seven Southwest states, is pumped through canals, to settling ponds, through filters into the lined growing ponds.

Clean fresh water and nutrients are added daily to feed the algae. No pesticides or herbicides are used.

Cultivating pure algae

Keeping out weed algae

Hundreds of aquatic organisms can bloom in nutrient rich water, as in lakes or swimming pools. Unlike a garden, weeding out algae is difficult since this algae is microscopic. Preventing unwanted algae is key to growing a pure culture.

Ecological pond management

Conventional farmers kill weeds and pests with pesticides and herbicides, leaving residues in the environment, on farm workers, and in food. Earthrise scientists keep out weed algae without toxic chemicals using a specially designed pond system and balancing pond ecology. Producing spirulina under these controlled conditions does not allow growth of contaminant or weed algae as in lakes and waterways.

Carbon Dioxide is bubbled into the water

Plant leaves take in carbon from the carbon dioxide (CO2) in the atmosphere. Algae grows so quickly that atmospheric CO2 cannot penetrate the water fast enough to sustain growth, so it is added. The CO2 used in carbonated drinking water is pumped into ponds.

Pumps send algae water to the sealed harvest building. Spirulina separated from water by series of screen and filters.

Pure mineral nutrients

Adding manure or organic matter directly into the water can foul the ponds and disturb algae growth. Instead, clean, pure sources of mineral nutrients are used. Minerals with nitrogen, potassium, iron and trace elements nourish high quality spirulina.

Continuous harvest

During the growing season, April through October, ponds are harvested every day. In the peak summer, harvesting occurs 24 hours a day, to keep up with explosive growth.

15 minute journey from ponds to dry powder

Spirulina flows through the stainless-steel harvest and drying system, never touched by humans. First screen filters out pond debris. Next screens harvest the microscopic algae. Nutrient rich water recycles back to the ponds. Final filter thickens spirulina from green yogurt to green dough.

Quick drying preserves nutrients

It is still 80% water inside the cells and needs drying immediately. Spirulina harvest is sprayed into the drying chamber to flash evaporate the water. Dry powder is exposed to heat for several seconds. Then it is vacuumed into a collection hopper. This quick process preserves heat sensitive nutrients, pigments and enzymes.

Shipped all over the world. Automated bottling line.

Sealed in oxygen barrier containers and bottled

No preservatives, stabilizers or additives are used, and the product is not irradiated. Powder has a high content of heat-sensitive phycocyanin, attesting to its quality. Dry spirulina can be stored for five years or more in special oxygen barrier containers, holding nearly full beta carotene potency. Powder is compressed into tablets and sealed into glass and plastic bottles for Earthrise® products.

DIC Earthrise phycocyanin extraction plant for Lina Blue.

Blue food color extraction plant

Spirulina has high phycocyanin, a natural blue pigment. In 2015 Earthrise Nutrtionals added a phycocyanin extraction plant to meet the rising demand for this natural blue colorant as major food companies shift from artificial to natural ingredients in the global food and beverage market.

Earthrise nutritional ponds and processing plant.

Quality control program

Technicians collect samples from pond water and dry product for dozens of analytical tests. Only after each lot has passed all tests is it certified ready to ship, accompanied by a lab report. Outside labs periodically confirm protein, pigments, amino acids, vitamins, minerals and microbiology. Earthrise developed a program to assure toxic algae are not present in spirulina ponds. Scientists in cooperation with university researchers developed immunoassay and enzyme inhibition bioassay methods to detect toxins.

ISO 9001 Certification

In 1998 Earthrise was ISO 9001 registered for the design, production, tableting and bottling of bulk algal products. ISO 9000 series of international quality standards are recognized in 80 countries. ISO 9001 is the most comprehensive and covers product design, engineering, manufacturing, purchasing, marketing and sales.

Meets food safety and quality guidelines

Earthrise is subject to inspections by the U.S. Food and Drug Administration. Earthrise Spirulina is GRAS (Generally Recognized as Safe) after FDA review of the scientific evidence on its safety, and has achieved third party HACCP (Hazard Analysis Critical Control Point) certificate of registration.

Lake Texcoco near Mexico City.

Worldwide farms

Three kinds of commercial farms operate today: 1. Lake farms harvest spirulina from natural lakes. 2. Outdoor pond cultivation with open ponds or covered greenhouses. 3. Photobioreactors.

Lake harvest farms

Lake harvesting enjoys advantages of available nutrients in the alkaline lakes, but quality may be inconsistent.

Mexico: In the 1970s, a Mexican company realized spirulina in Lake Texcoco was clogging extraction of soda ash from the lake. The world's first large plant was built here. In 1979 Mexican spirulina was the first to be exported to the U.S. for health food products. Spirulina Mexicana has been closed since the mid-1990s.

Twyn Taung Lake, Myanmar.[2]

Myanmar: In 1988, commercial harvest began on alkaline volcanic lakes that enjoy natural blooms of spirulina. By 1999 production increased to 100 tons per years and sold in the national market. Cultivation ponds grow more spirulina alongside Twyn Taung lake.

Chad: Alkaline lakes around Lake Chad in Africa offer an ideal location. Indigenous Kanembu women harvest spirulina manually from the water and sell sun dried *dihé* in the local markets, distributing over 200 tons per year.

Cyanotech on the Kona Coast, Hawaii.[3]

Outdoor pond cultivation systems

Most commercial farms are designed and built with shallow raceway ponds circulated by paddlewheels. Ponds vary in size up to 5000 square meters, and water depth is usually 15 to 25 centimeters. They require more capital investment than lake farms, and operate at higher efficiency and quality control.

Hawaii, USA: Cyanotech opened in 1985 on the Big Island of Hawaii and produces 400 tons of spirulina per year. It is ISO 9001 Certified and Generally Recognized as Safe (GRAS) by the US FDA for all food, beverage and supplement applications. Cyanotech also produces haematococcus for astaxanthin.

Boonsom Spirulina Farm Chiang Mai Thailand.[4]

Thailand: Boonsom Farm near Chiang Mai is a medium size (40,000m2) family-owned spirulina farm. It has produced finished products for the regional market in Thailand and Asian countries for the past 20 years. Boonsom Farm is promoted as an algae-tourism destination, offering a tour of the facilities, samples of spirulina foods and a spirulina health spa.

India: Research began in late 1970s, from family scale to production farms. In 1990 India set a national standard for spirulina. Parry Nutraceuticals began spirulina production in Tamilnadu in 1996 and expanded into haematococcus in 2003.

Parry Nutraceutical spirulina ponds in India.[5]

Bluetec spirulina in Inner Mongolia China.[6]

China: China is now by far the world's largest spirulina producer with numerous farms across China from Hainan Island in the south to Inner Mongolia in the north. With an annual capacity in the thousands of tons, China is now the major world exporter.

Other farms: There is commercial production in Australia, Myanmar, Taiwan, Cuba, Chile, Vietnam, Israel, Bangladesh, Philippines, Martinique, Peru, Brazil, Spain, Portugal, Australia and other countries. Spirulina farms are multiplying around the world.

Australian Spirulina in Darwin.[7]

Jongerius Ecoduna photobioreactor plant. Austria.[8]

Photobioreactors

Spirulina needs hot temperatures and most temperate climates are too cold for outdoor pond cultivation year round. This limits where it can be grown economically. To grow high value algae and in colder climates, companies have developed photobioreactors.

Indoor bioreactors use tanks and artificial lights, growing algae like bacteria and yeast. Outdoor bioreactors are designed in modules of vertical cylinders or rows of transparent tubes along the ground. Plexiglass tubes and coils act as solar collectors, increasing temperature and extending the growing season.

Advantages are higher productivity and culture density than open ponds, less water evaporation, greater culture control, and climate independence, for productions in cold climates.

Disadvantages are higher capital costs than ponds, higher operating costs for lighting. Algae may stick to the inside of tubes and block sunlight or get too hot. Excessive oxygen produced by the algae while growing can reduce growth. When scaled up for commercial production, these systems may not compete with lower cost open ponds, but are useful for high value algae.

Companies use photobioreactors for algae such as spirulina, chlorella, haematococcus, nannochloropsis and others for food, pharma, industrial, cosmetic and aquaculture products.

AgCore spirulina farm. Rhode Island USA.[9]

Integrated farm with CO2 capture technology

Agcore Technologies vertically integrated spirulina farms operate year round with daily harvesting, processing and packaging. Finished products include vegan snack foods and smoothie powders, fish food for the aquarium hobby industry and aquaculture.

A designed modular CO2 capture system can separate carbon dioxide from waste flue emissions from power plants, fermentation plants and cement producers. Then a process to infuse CO2 in water without creating bubbles provides high gas utilization efficiency. The CO2 saturated water is good for higher yields of algae.

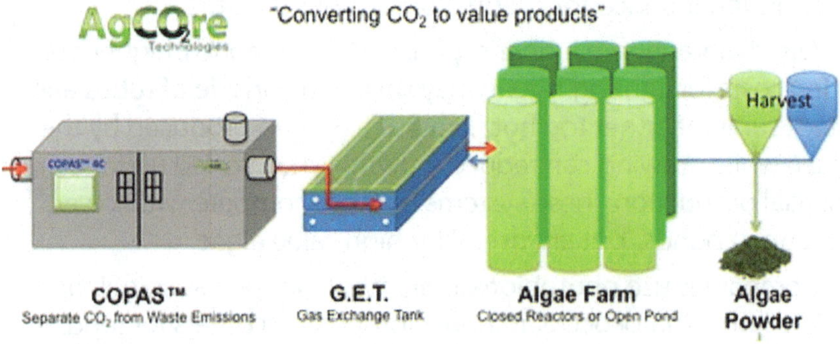

Design of the modular CO2 capture system to grow algae.

Spirulina La Capitelle microfarm in France.[10]
Philippe Calamand, Ripley Fox, Denis von der Weid. 2002.

Spirulina microfarmers in France

Parallel to large commercial farms has been the evolution of village farms with appropriate technology in the developing world in the 1980s and 1990s. Many technicians working for NGOs originated from France, and returning workers became the first algaepreneurs to build greenhouse microfarms there beginning around 2002 offering spirulina for local markets.

Pioneers Ripley Fox and Jean-Paul Jourdan catalyzed the movement to grow spirulina in France. Today there are 160 members of the Federation des Spiruliniers de France, offering courses and curriculum for growing spirulina. These microfarmers have developed many innovative fresh, frozen and dried spirulina products.

Federation des Spiruliniers de France has 160 members.[11]

Eco-domaine microfarm in Normandy 2011.[12]
Laurent Lecesve and Gilles Planchon.

Experimental integrated algae microfarm

An integrated organic spirulina microfarm in the far north of France, 3 km from the Normandy Coast, Eco-Domaine Ferme de Bouquetot, has a series of algae ponds.

Ponds are covered, insulated and heated to retain warmth to extend the short growing season. An experimental biogas digester converts farm waste into biogas for heating ponds and digested organic nutrients for feeding algae. Eco-Domaine offers training programs for spirulina producers.

Eco-domaine spirulina based lunch with Jean Paul Jourdan 2011.

Greenhouse of La Spiruline de Haute Saintonge, France.[13]

Algae microfarms are proliferating

A common request over the past 40 years comes from people who want to grow algae at home or in their community. Microfarm entrepreneurs are springing up around the world. There are now 160 spirulina microfarmers in Europe and trainings for growing algae. Growers are producing their own products and selling directly in their local region. Remote sensing apps will assist culture monitoring and diagnosis. On a small area, a community could meet a portion of its food needs from microalgae, freeing cropland for community recreation or reforestation.

Haute Saintonge: Harvesting and dewatering spirulina. Extruding noodles on drying shelves, dry product.

***Energaia rooftop photobioreactors in Bangkok Thailand.*[14]**

Urban and rooftop systems

Spirulina can be grown in underutilized urban and suburban areas and even rooftops. Energaia began production in Bangkok Thailand by making a deal with the Novotel Bangkok on Siam Square to grow spirulina on the roof and sell it in the building cafe. The company developed circular photobioreactors, sells fresh and dried product, spirulina pasta and has expanded to Indonesia, India, Bangladesh and Vietnam.

Spirulina is ideal for growing in small spaces, at home, in yards at schools or parking lots for family, friend and community consumption. Do-It-Yourself aquarium kits and educational courses are available online to get started.

The author's home growing systems: indoor tanks 2012, outdoor deck 2013, rooftop 2019.

Spirulina Viva greenhouse.[15] *2015.*

Spirulina Viva Mexico

Algaepreneurs are starting microfarms in North, Central and South America. Spirulina is native to Mexico and was an original food of the Aztecs. Spirulina Viva in San Miguel de Allende is the first family owned and operated algae microfarm in Mexico.

Here, on their homestead, Katie and Paco with their boys, grow and sell fresh and dried spirulina products and healthy spirulina food snacks direct to the local community in San Miguel, through retail health shops and by express mail all across Mexico.

Spirulina Viva offers tours and training programs for prospective algae entrepreneurs, reviewing all aspects of the microfarm, greenhouse cultivation, harvesting, pressing and packaging. Tasting popular foods with spirulina and then taking home manuals and bottles of seed culture to get started.

Spirulina Viva fresh spirulina squares and snack foods.

Smart Microfarm greenhouse webcam view. Summer and winter. Northern California 2018.[16]

Smart Microfarms in California

Smart Microfarms has operated a research station in Richmond California since 2012 and a microfarm near Olympia Washington. The NorCal farm opened in a former flower greenhouse in 2016, testing smart apps and web-based technology to send real time data so a remote algae expert can monitor and guide a local microfarmer.

SpiruSource fresh, frozen and dry spirulina is sold online, direct-to-consumer, at farmers markets and retail stores in Washington State, San Francisco Bay Area, Santa Cruz and Los Angeles.

SpiruSource poster- fresh, frozen, dry products, 2018.

Chapter 7

Spirulina in the developing world

- **Benefits of one tablespoon a day**
- **Bioneering visions**
- **Integrated village system experimental projects**
- **Child nutrition and small scale farms in India**
- **Evolution of spirulina production in Chad**
- **Antenna Technologies in Africa and Asia**
- **Spirulina in india and Myanmar**
- **Hope for the world's children**

One dream behind spirulina was developing it as a new food resource for children in a hungry world. One tablespoon a day offers remarkable health benefits.

The UN World Health Organization (WHO) in Geneva confirmed: "Spirulina represents an interesting food for multiple reasons, and it is able to be administered to children without any risk. We at WHO consider it a very suitable food."[1]

Benefits of one tablespoon a day

Spirulina offers health benefits to undernourished people. Beta carotene can overcome eye problems caused by Vitamin A deficiency. One tablespoon a day can eliminate iron anemia, the most common mineral deficiency. It is the most digestible protein food, important for malnourished people whose intestines can no longer absorb nutrients effectively. These benefits have helped rapid recovery of children from malnutrition related diseases in Mexico, Togo, Romania, China, Rwanda, Zaire, India, Ukraine and Belarus.

Bioneering visions

In the 1960s and 1970s, small groups of scientists and visionaries understood what microalgae could become. Larry Switzer, wrote about his hope for a breakthrough in production:

"It had to be more productive than conventional agriculture, adaptable to different climates and cultures, appropriate ecologically, economically and socially, independent of the vested interests in world food production and distribution, capable of relying on renewable energy and waste or abundant raw material resources. It would have to represent a major expansion of the photosynthetic energy base that supports all life on Earth. Finally, it would have to radically improve the supply, distribution and consumption of protein to millions of pregnant and nursing mothers and children."

Larry Switzer • Denise and Ripley Fox

"It is critical to provide nutrition to deprived embryos and infants to preserve the precious creative genius that is waiting to be released from each fully developed human mind."[2]

In 1976 Switzer proposed the Nigerian Government use oil money to finance farms on the shores of Lake Chad to feed the country's population. But the novelty of algae, lack of interest in food, and a political coup doomed this plan. The lesson was to develop technology and a market in the U.S. first, then apply it to the Third World. Switzer founded Proteus Corp. and built the first farm in California. Earthrise Company introduced spirulina in 1979.

Beginning of village scale technology

Parallel to commercial production was the work of Ripley and Denise Fox, who founded the non-profit *Association Pour Combattre la Malnutrition par Algoculture* (ACMA).[3] They grappled with a fundamental problem: lack of opportunity. Commercial farms could produce spirulina, but if the hungriest people had no money to buy it, they would never get any. They must grow it themselves, but how could a village project be self-sustaining? The first step in improving health is eliminating health problems through sanitation and waste treatment. "One recycles the wastes already present in the village," says Dr. Fox in his books, *Algoculture: Spirulina, Hope for a Hungry World,* and *Spirulina, Production and Potential (1996).*[4]

The integrated health and energy system

Enthusiasm, participation and training of village people are vital to show how they can improve sanitation, health, nutrition and generate income from compost, biogas fuel, fish and algae.

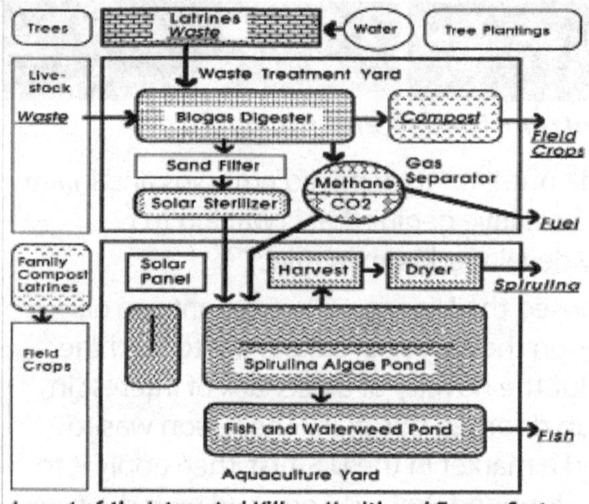

Layout of the Integrated Village Health and Energy System
(R. Henrikson 1994)

The Integrated Village Health and Energy System design won the prestigious 1987 *European Award for Appropriate Environmental Technology*, sponsored by the EEC and the United Nations Environmental Program.

Latrines eliminate the source of intestinal parasites. Family latrines can produce compost to improve soil fertility. In a village center, central latrines provide wastes to a biogas digester.

The digester ferments animal and plant waste into gas, liquids and solids. A gas separator splits biogas into methane for cooking and lighting and CO2 for spirulina. Liquid effluent is sterilized for-nutrients for spirulina. Sludge is composted and spread on soils.

Solar panels run pond paddlewheels, auxiliary lighting and recharging batteries. Spirulina is screened from pond water and added to the pond to feed fish or dried in a solar drier for food.

This is a systemic approach to community problems. Each successful project becomes a training center for surrounding villages.

Dr. Ripley Fox discussing cultivation in Karla, India. 1980s.

"Every nation is supported on the shoulders of its villages," says Fox. "We believe by providing technical assistance to improve sanitation and agricultural output and by saving the trees, we can increase the vitality of these villages."[5]

Karla, India – the first integrated system

Karla village is near Wardha, famous for Mahatma Gandhi's ashram. The Center of Science for Villages (CSV), inspired by Gandhi's vision that India's strength lies in the strength of its villages, operated the system. This project derived income from compost and fish sold to local villagers, and from spirulina sold in Bombay. The CSV distributed spirulina cookies and noodles to local children, and spirulina chapatis were another local favorite.

Since algae grown on waste nutrients must be proven safe to consume, it was important to show pathogens in human and animal waste were destroyed in the biogas digester and solar sterilizer. A six month study showed pathogens were virtually eliminated and were lower than in other local foods.

Tending the growing basin. Harvesting spirulina through screens. Farende, Togo, 1989.

Farende, Togo – a remote African village

Less than 100 years ago, Northern Togo was home to lush forests. Today little stands in the way of the Sahara sweeping south. These arid grasslands are home to an exploding human population. The Kabyé people grow millet and root crops. But soil quality in marginal. Growing numbers of people stress the carrying capacity of the environment.

The village of Farende participated in an experimental project to grow spirulina. A small 100 square meter basin could grow enough to supplement the diet of 100 children a day. Pouring pond water through a screen, spirulina became a thick paste, then loaded on a screen in a solar dryer and distributed at the health clinic.

Undernourished children took spirulina as a daily supplement at the clinic. The nurse told mothers about its benefits. One tablespoon a day with water brought remarkable results. Children found the taste of this "green medicine" acceptable and within a week showed improvement and weight gain. Mothers brought children to participate in a feeding study.

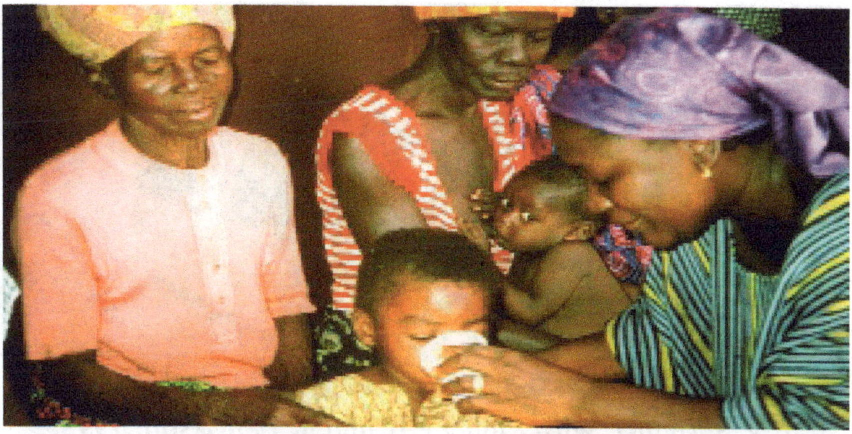

Clinic nurse introduced children to spirulina. Farende 1989.

From 1984 to 1990, the Foxes assisted Eglise Evangelique du Togo building this experimental system, operated by young village men. Also cooperating was the U.S. Peace Corps, teaching soil composting, vegetable gardening, conservation and growing trees. The first corporate sponsors were DIC of Japan and Earthrise Company of California.

People of Farende, Togo (Collage by R. Henrikson).

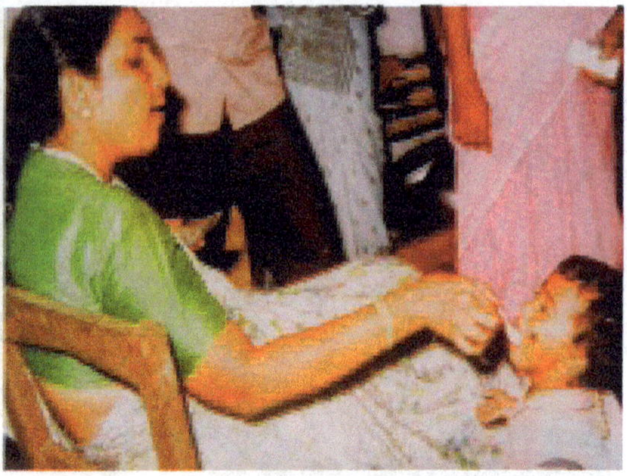

C.V. Seshadri, with Ripley Fox and R. Henrikson, 1993.
A spoonful of "Spiru-Om" is given to a child. C.V. Seshadri.[6]

Family scale cultivation in India

The *All India Coordinated Project on Algae* began in 1976 to harness algae as biofertilizer, human nutrition and animal feed at commercial and rural levels. Original work began with scenedesmus, green algae, but later, spirulina was chosen because of its advantages. In 1991, the Indian government launched large scale nutritional studies and issued official standards for food grade spirulina.

Reducing eye disease

In 1992, in the world's largest spirulina nutrition program, 5,000 children near Madras consumed one gram a day for 150 days, providing the daily requirement of beta carotene (Vitamin A) which helps prevent eye disease. A symptom, *Bitot's spot,* scarring of the conjunctiva of the eye, decreased from 80% to 10%.[7]

Spirulina was given to children in noodles, sweetened with sugar to preserve beta carotene. *"Spiru-Om,"* was well accepted by the children. This project, sponsored by the Indian government, was lead by Dr. C.V. Seshadri of the Murugappa Chettiar Research Center in Madras.

Chad traditional: Spirulina is dried in sandy hollows.

Improved: harvesting through fine cloths, extruding noodles, solar drying and packaging. By Mahamat Sorto, FAO Chad, 2011[8].

Evolution of production in Chad

Kanembu women have harvested *spirulina* from lake regions near lake Chad using traditional methods for centuries. About 1600 ladies harvest from small natural alkaline soda lakes, and produce over 200 tons of *dihé* per year. Improvements in spirulina harvesting through filter cloths and dehydrating in solar dryers was introduced, raising the income of women. This project was funded by the EU and implemented by FAO.

Spirulina greenhouse in Chad. 2008. Photos: Georges Bonnin.

Antenna Technologies in Africa and Asia

Antenna was established in 1989 by Denis von der Weid, promoting spirulina against malnutrition with a mission to make algae more affordable. Today Antenna Foundation in Switzerland (antenna.ch)[9] is engaged in research and dissemination of technologies for the needs of the poor in developing countries.

Antenna France (antenna-france.org)[10] has been working to combat malnutrition since 2002 and works with local partners in Africa and Asia to construct independent and financially viable spirulina farms and distribution networks. Antenna also manages nutrition centers and conducts research and testing of food products enriched with spirulina.

Antenna France circular pond was awarded First Place Prize in the International Algae Competition for algae production.

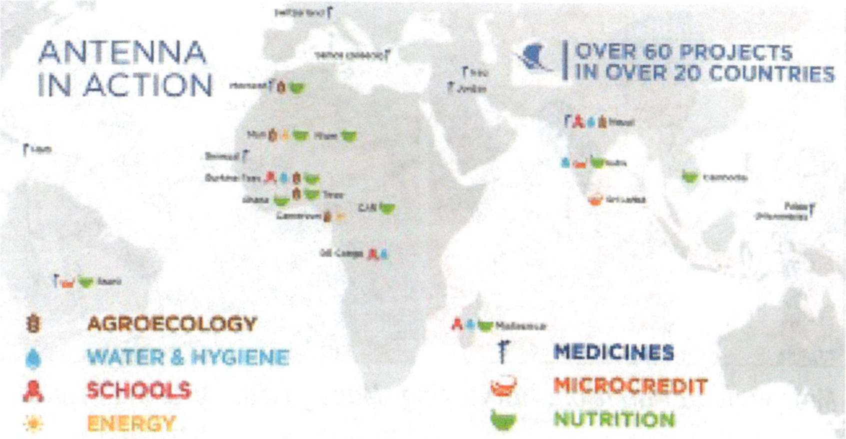

Antenna Technologies in Africa and Asia

Antenna farms are in Burkina Faso, Cambodia, Madagascar, Laos, Mali, Mauritania, Niger, Togo, Central African Republic and India.

Togo

This farm, about 150 km from the capital Lomé, was established in 2004. A portion of the product has been distributed to mothers and children through the Research Center of Nutrition and other clinics. Objectives are expanding social distribution to provide dietary supplements to 3000 people each year, and ensuring financial autonomy of the project.

Agou Nyogbo farm, Togo. Antenna-France.

Maison de la Nutrition. Harvesting. Madagascar. Antenna-France.

Madagascar

Antenna has been building farms since 2005. In 2010, Antenna opened Maison de la Nutrition in Antsirabé to distribute a daily diet with added spirulina and educate mothers and children about balanced nutrition and elementary hygiene.

Antenna Spirulina Nutritech in India

The first farm was launched in 1995. Technical support from Antenna Switzerland helped to enlarge the farm. Over the years, Antenna has trained people from non-governmental organizations (NGOs) to grow spirulina in small farms all over India and Bangladesh. To increase the number of people consuming spirulina, a marketing division developed new products.

Antenna Nutritech spirulina ponds, products and programs for children. Spirulina Nutritech Foundation.[11]

In Cambodia, Siem Reap and Kandal farms.

Cambodia

Since 2011, Antenna partnership farms near Angkor Wat and Phnom-Penh are run by local families growing mushrooms and vegetables. Spirulina has been approved by the Ministry of Health for spirulina sales and distribution in Cambodia. Two-thirds is commercially marketed and one-third is distributed through humanitarian organizations to children.

Burkina Faso has nine farms

The first farm was launched in Koudougou in 1999. By 2001, success led to expanding existing farms and setting up new ones with the assistance of NGOs to meet growing demand. Burkina Faso now has 9 farms or more. Most are members of a national spirulina association and use standardized packaging.

Spiruline du Burkina. • Nayalgué farm in Koudougou.

Aurospiral farms in Auroville India[12]

Beginning in 1978 with chlorella and scenedesmus, the Auroville algae farm shifted to spirulina in the 1990s. Since 1997, Aurovilian Hendrik van Poederooijen from the Netherlands has operated the farm and expanded operations.

Today Aurospiral is growing spirulina by simple eco-friendly methods at two locations in Auroville. Around four tons a year is mostly sold within the Indian domestic market. Aurospirul has established itself as a well-known brand of spirulina.

1978 algae farm in Auroville • 1998 Hendrik's first harvest.

Harvesting with boats on Twyn Taung Lake. (Min Thein[9])

Producing spirulina in Myanmar[13]

Four volcanic lakes have natural spirulina blooms. Production began at Twin Taung Lake in 1988, and by 1999 increased to 100 tons per year. About 60% was harvested from boats on the surface of the lake, and 40% was grown in ponds alongside the lake. In recent years, lakes have reduced algae blooms, as the lake ecology has changed because of nearby development.[9]

At June Pharmaceuticals farm near Yangon, spirulina is grown in ponds, harvested on inclined filters, detwatered and pressed. This paste is extruded into noodles, dried in the sun on sheets, taken to a factory, and made into a wide range of products including tablets, health supplements, snacks, foods, soups, beverages, cosmetics, shampoo, biofertilizer and even spirulina beer. Almost the entire production is distributed inside the country.

Pond harvest with screens and sun drying. 2013.

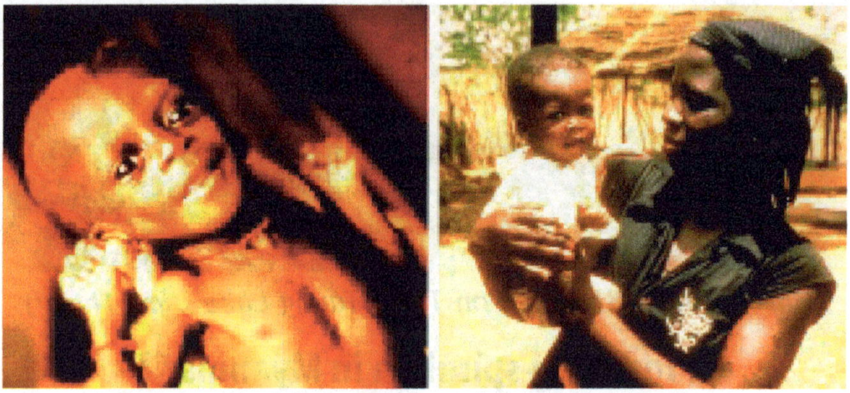

Baby saved from malnutrition by spirulina, before and after in Togo, West Africa.

Hope for the world's children

Spirulina has been recognized as a "national food" by the two most populous countries, India and China, where cropland is in short supply. Both countries are becoming major world producers. In India and Vietnam, it is recommended for nursing mothers. For newborns, malnutrition is often caused by a lack of mother's milk, the mother herself often being ill. Spirulina given to the mother helps a return to lactation and the babies rapidly gain weight.[14]

In Ho Chi Minh City, Vietnam doctors conducted nutrition trials in two orphanages. Food formulas containing 5% spirulina had a better effect than soya at much lower levels.[15]

In a bush hospital in Zaire in 1990, spirulina in corn flower cookies improved health of children with malnutrition.[16] In Central African Republic in 1992, a clinic treated 200 children a day, improving health of those with kwashiorkor and marasmus.[17] In Rwanda in 1993, children with kwashiorkor were given algae. After 15 days, their mothers wanted to buy spirulina and learn how to grow it.

Mothers and children around the world are embracing spirulina. We hope to witness even more widespread acceptance.

Chapter 8

Algae's impact on food, agriculture and climate

- **Problems facing global food production**
- **Hidden costs of food production**
- **Spirulina advantages: land, water, energy**
- **Spirulina compared to other microalgae**
- **Advantages of algae production**
- **Coming intergrated microalgae biorefineries**

Conventional food production hides environmental costs. The cost accounting system doesn't take nature into account. You pay for externalized costs, but not at the cash register.

Spirulina has no hidden costs. Spirulina delivers more nutrition per acre than any other food, conserves land and soil and uses water and energy more efficiently than other foods. As global algae production expands using non-fertile land and brackish water, cropland can be returned to forest. As people eat lower on the food chain, we can help regreen our planet.

G

Three problems facing global food production

Agricultural: Limits of the Green Revolution

Many consider the green revolution of the 1960s and 70s to be a successful achievement in the developing world. But success is not distributed evenly. New seeds, fertilizers and pesticides boosted export crops of wealthier farmers. Yields of locally consumed food of subsistence farmers on marginal land did not benefit as much.

We are nearly 8 billion people. As people grow wealthier, there is demand for more foods such as meat and dairy which require large amounts of grains. Unfortunately, grain growing area peaked in 1981, and has fallen more than 5% since. Yields are down as irrigated area has begun to shrink and fertilizer use is curtailed due to diminishing returns. Big losses of fertile farmland result from desertification and soil depletion. We are running out of space.

World Resources Institute agrees new approaches are needed. "Agricultural research has changed since the days of the Green Revolution... some researchers are increasing emphasis on needs of poor farmers and on ecologically sustainable agriculture."[1]

Political: Food is not evenly distributed

Many food experts claim the problem is not production, it is equitable distribution. Although the world food supply is adequate to end hunger, chronic hunger persists, victimizing one third of the world's people, and primarily children. The underlying cause of political instability in these regions is the inability of people to control their own resources. Hunger persists wherever people lack opportunity to participate in their society.

Mega drought forecast for US Western food growing regions?

Environmental: Food growing areas are declining

Desertification and soil depletion worsen from the stress of chemically produced export crops, livestock overgrazing, poor soil management and rainforest destruction.

A 1983 U.N. Food and Agriculture Organization (FAO) report concluded the developing world is capable of producing sufficient foods to sustain its own population.[2] Yet, because unrestricted movement of food in the developing world is unrealistic, 65 countries have insufficient resources to meet food needs. These regions include over 20% of Africa where millions of people live.

Worldwatch reports desertification claims 15 million acres worldwide each year. According to the U.N. Environmental Program (UNEP), "35% of the earth's land surface is threatened by desertification and, with them, one-fifth of humanity."[3]

All three problems combine for maximum impact: 1) The green revolution has reached its limits, 2) world food distribution is not likely to change dramatically, and 3) cropland is lost due to economic growth and environmental deterioration. As farmed land turns into desert, or paved over for infrastructure, less fertile land is available while population is growing. To face these problems we need unconventional food sources with higher yields.

Rainforest destruction in Borneo.

Hidden costs of food production

Agribusiness farming practices have externalized many production costs. You still pay for these costs, but not at the checkout counter. If you add hidden costs of food you pay indirectly, food would be much more costly.

What are the hidden costs and how did they arise?

1. Medical costs from unhealthy food. Pesticides, fungicides, animal antibiotics, preservatives, additives and fatty, salty foods create long term health risks. Over processed foods represent an unhealthy diet and higher medical bills, insurance and taxes. Health care is the fastest growing sector of the economy.

2. Farm subsidies. Taxes for government farm subsidies support agribusiness and encourage wasteful consumption of water and soil. Hidden subsidies encourage waste and make food appear to be cheaper to produce and cheaper to buy.

3. Toxic cleanup costs. Pesticides, herbicides and chemical fertilizers pollute our water and land. We will pay much more tomorrow for cheap food today.

4. High global military costs. Developing countries need cash to pay interest on debt to our banks. They produce food to export, taking land away from local production and displacing small farmers. This creates food insecurity and chronically hungry people. One response has been militarization to maintain security.

5. Government debt and interest costs. Hidden costs in increased health services, military spending and farm subsidies gobble up government revenues. Budget deficits and interest costs pull money away from investment in productive assets, blocking meaningful toxic cleanup and environmental restoration.

6. Environment and resource destruction. Agribusiness treats fertile soil and precious fresh water like factory assets to be depreciated. Natural wealth is extracted, and not replaced. It accounts for soil demineralization, salinization and erosion, the drawdown of water aquifers, and the loss of forests world.

"The burger that ate a rain forest". London Times 1989.

Does a fast food 1/4 pound burger cost only $3.99?

For American grain fed beef, it takes16 times as much corn to get protein from beef than from corn directly. Each pound of corn causes 2 pounds of topsoil erosion. Each 1/4 pound burger costs 8 pounds of topsoil.[2] Soil and water alone exceed the $3.99 price!

An imported beef burger takes 55 square feet of tropical forest cleared for grazing land. Expanding cattle ranches have caused massive rainforest destruction in Brazil. Deforestation releases about a billion tons of carbon into the atmosphere each year.

This burger could really cost $100, depending on how one values Earth's resources. If everyone had an American appetite, the entire planet would become one giant beef farm.

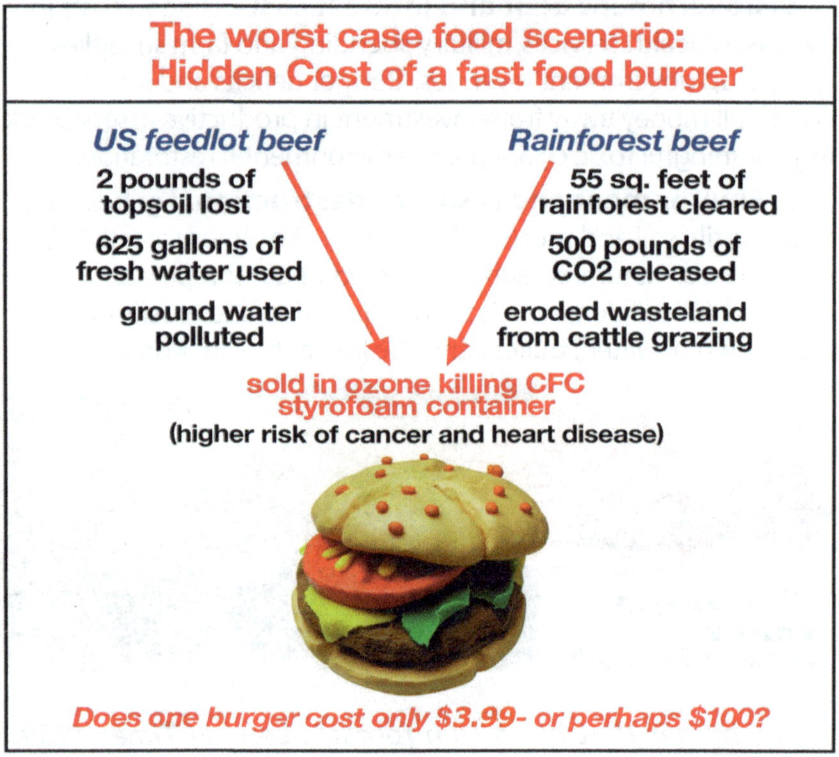

© 1994 R. Henrikson

7. Accounting doesn't take nature into account.

We do not know what a fast food burger really costs because the economic system ignores natural resource depletion. Gross National Product (GNP) figures and company balance sheets show man-made capital depreciation, but not consumption of soil, water, trees, fisheries or wildlife.

Accounting methods evolved years ago when natural resources were considered free and unlimited. We need to begin taking nature into account. Only when the world economy shifts to natural resource accounting will we be able to measure the true cost of our food and all other products.[4]

Spirulina prices reflect true costs

Growing spirulina does not hide costs. Eating spirulina will improve health and lower medical bills, compared to a diet rich in meat and conventional foods. There are no big government subsidies. Ecological cultivation does not cause pollution, soil erosion, water contamination or forest destruction. Spirulina will become more competitive with conventional proteins when hidden costs are taken into account.

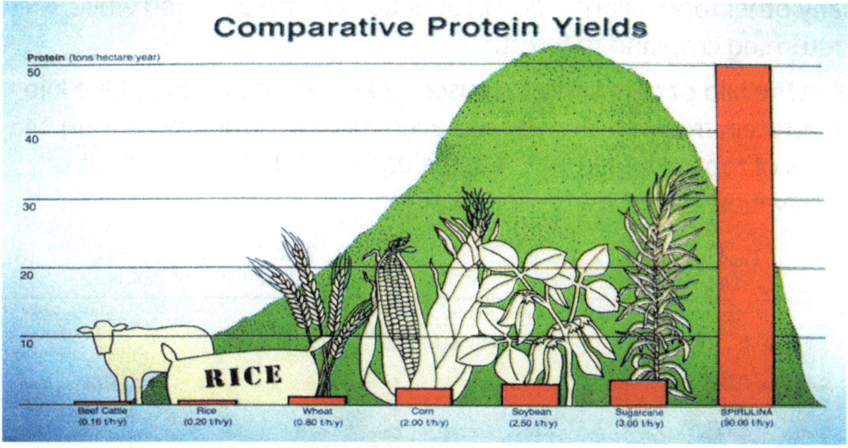

Spirulina has 20 to 200 times protein yield as other foods. (The Futurist Feb.1985).

A big CO2 consumer and oxygen producer

Forests help absorb carbon dioxide. Trees are the best land plants for fixing carbon, from 1 to 4 tons per hectare per year.[4] In the California desert, spirulina fixes 6.3 tons of carbon per hectare per year and produces 16.8 tons of oxygen. In the tropics it is 2.5 times more productive.[5] Carbon budgets will be taken seriously as our planet struggles with global warming from the buildup of atmospheric CO2.

Spirulina resource advantages
An environmentally sound green food revolution

Land and soil are conserved

Spirulina is 60% protein and can be cultivated on marginal and non-fertile land. Rapid growth means spirulina protein needs 20 times less land than soybeans, 40 times less than corn, and 200 times less than beef. Spirulina offers more nutrition per acre than any other food. Higher food production can be achieved while returning cropland to forest.

One kilo of corn protein causes 22 kilos of topsoil loss. One kilo of protein from corn-fed beef is even more destructive, causing 145 kilos of topsoil loss from the cattle eating all that corn. Spirulina cultivation causes no topsoil erosion.

Land Area Needed to Produce One Kilogram of Protein

	Sq. Meters	Quality
Spirulina[a] 65% protein	0.6	non-fertile
Soybeans[b] 34% protein	16	fertile
Corn[b] 9% protein	22	fertile
Grain-fed Feedlot Beef[b] 20% protein	190	fertile

[a] Y. Ota, Earthrise Farms, California 1995
[b] Leesley, et al. "A low energy method of manufacturing high-grade protein using spirulina," University of Texas, 1980. Pimentel, 1975, USDA

Water Needed to Produce One Kilogram of Protein

	Liters	Quality
Spirulina[a] 65% protein	2100	brackish
Soybeans[b] 34% protein	9000	fresh
Corn[b] 9% protein	12500	fresh
Grain-fed Feedlot Beef[b] 20% protein	105000	fresh

[a] Y. Ota, Earthrise Farms, California 1995
[b] *Diet for a Small Planet*, 1982, pg. 76-77, Dr. David Pimentel, Cornell University, 1981.

More efficient water use

Even though spirulina grows in water, it uses far less water per kilo of protein than other common foods. At spirulina farms, water is recycled back to the ponds after harvesting. Production ponds are usually sealed with liners and very little water seeps through the ground compared to conventional crops. The only significant loss is by evaporation.

Spirulina protein uses 1/3 the water as soy, 1/5 as corn, and 1/50 the water needed for beef protein. Fresh water is one of the world's most critical resources. Spirulina can use brackish or alkaline water, unsuitable for agriculture. Growing algae for food will become more attractive since it does not need to compete with water for drinking or agriculture.

Energy Efficiency
(Million Kjoules Per Kilogram of Product)

	Total Energy Output	Food + Residual Energy Output	Energy Output/ Input
Spirulina[a] 65% protein	3.8[b]	23	6.1
Soybeans[b] 34% protein	11.7	13.8	1.2
Corn[b] 9% protein	5.5	16.5	3.0
Grain-fed Feedlot Beef[b] 20% protein	456	16	.04

[a] Y.Ota, Earthrise Farms, California 1995
[b] Leesley, et al. "A low energy method of manufacturing high-grade protein using spirulina Universitiy" of Texas, 1980, Pimentel, 1975. USDA

More efficient energy use

Spirulina requires less energy per kilo than soy, corn or beef, including solar and generated energy. Its energy efficiency (food energy output per kg / energy input per kg) is 3.5 times higher than soy, 1.4 times corn, and 100 times grain fed beef.

Comparing spirulina to other microalgae

Other microalgae sold as food supplements are chlorella (green algae), aphanizomenon flos-aquae (blue-green algae), dunaliella (red algae), haematococcus (green algae) and nannochloropsis.

Spirulina has been most commonly adopted of all the microalgae. There are large commercial farms and now many microfarms in 40 countries. Why spirulina? It has these six advantages:

1. Traditionally consumed and proven safe.
2. Easy to grow.
3. Easy to harvest.
4. Existing scientific research on health and medical benefits.
5. Established global market opened by commercial farms.
6. Low cost of entry for small scale production. There has been a proliferation of small scale spirulina farms aound the world in the past two decades.

Microfarms: Sekong, Laos (Antenna). Chad (George Bonin).

Advantages of microalgae production

For food self-sufficiency, arid areas need new food production and economic opportunity while simultaneously restoring the environment. Unconventional foods such as spirulina microalgae, aquaculture, salt-tolerant crops and drought-resistant grains and legumes offer help for arid regions.

Facing a serious loss of cropland and higher food imports, China declared spirulina a national food priority thirty years ago. Since then, over 50 spirulina farms have sprung up, many in southern China, making China the world's largest producer.

Areas of chronic malnutrition are most common in the arid tropics and subtropics. Here, diets are high in carbohydrates and sugars, but low in protein, certain vitamins and minerals. Spirulina thrives best in these locations, and is a perfect complement to the typical diet. It speeds recovery from malnutrition. It is more digestible than other plant, dairy or meat products. This is important for victims of malnutrition with poor ability to digest food.

Moving forward, it is likely a portfolio of microalgae will utilized, depending on climate, resources and desired output. Beyond spirulina, chlorella, nannochloropsis and others, scientists are investigating new algae strains for scaling up.

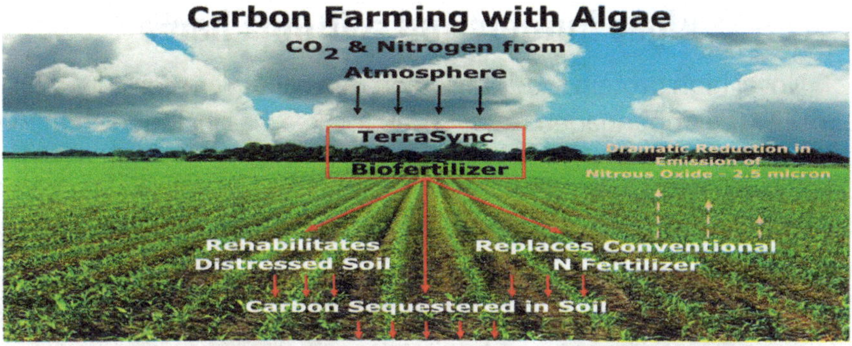

Accelergy Presentation (*Mark Allen*).[6]

Microalgae for regenerative agriculture

What does sustainable agriculture mean for the global agricultural system? One third of the world's topsoil is worn out and degraded, biodiversity is rapidly declining, and agriculture accounts for 24% of global carbon emissions. Going beyond merely sustaining, regenerative agriculture aims to renew soils and add diversity through self-nourishing ecoogical systems based on nature.

Regenerative practices improve soil health by using cover crops to retain moisture and build microorganisms, no-tilling to limit disruption of soil structure, crop diversity for more resiliant landscape, and integrating livestock to brake up compacted soil and add manure. Regenerative agriculture land will sequester more carbon.

Microalgae can help regenerative ag in several ways. Because algae protein yield is 20 to 200 times greater per area than crops or livestock, global protein need can be partially met. Algae protein grown on non-fertile land can free lands for regeneration or reforestation. Finally, microalgae biofertilizers on soils can rapidly rebuild soil structure for higher yields and more nutritious foods. Farmers can buy algae biofertilizer treatments or cultivate local algae, concentrate the culture and flow the nutrient rich biomass directly on the field for better yields and soils.

Algae burgers and spirulina heme

Novel high protein foods based on microalgae are coming. Space 10, a Copengagen research lab for Ikea, has included spirulina in their fast food of the future, the Dogless Hotdog- in healthy dark green spirulina buns.

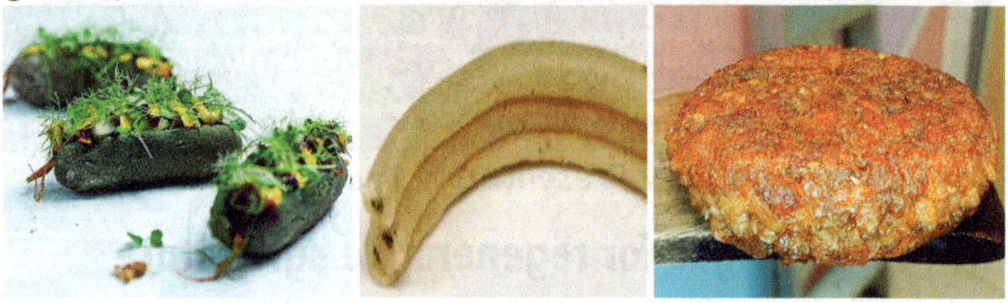

Dogless Hotdog spirulina bun[7] • 3D printed snacks • Algae burger

But algae's intense green color is usually a turn-off since people don't want green baked foods. Scientists at Wageningen Univ. in the Netherlands and Spanish Univ. in Valensia developed 3D printing for microalgae based cereal snacks, making foods more visually appealing by customizing shape, texture and color. Using a coaxial 3D printer, green color is "hidden" inside the healthy treats.[8]

The world's first burger from microalgae arrived in 2021 from Sophie's Bionutrients of Singapore. [9] It's made of a various microalgae including chlorella with a higher protein content than beef, showcasing the versatility of microalgae protein that's good for the planet.

What makes the plant based Impossible Burger taste like beef is a 'bleeding' heme analog produced through an insertion of DNA from soy plants into genetically engineered yeast. This yeast is fermented to produce heme. In 2021, A US company Back of the Yards announced a heme analog from spirulina phycocyanin extract with a natural umami and meaty taste that can challenge genetically engineered heme.[10]

Algae biofeeds for aquaculture, farmed animals

Plant based meat and fish products and alt-proteins are projected to gain wide acceptance this decade. Many meat substitutes like Impossible and Beyond burgers are entering the market, offering great tasting and healthier products with environmental benefits.

Meanwhile, billions of people will still be eating animal meat and fish protein. Algae supplements in aquaculture and animal feeds, replacing a postion of grains or fishmeal, can support healthier animals that have higher survival rates, need less antibiotics, grow faster, create less waste, provide superior nutrition and taste better.

The fish farming industry has quickly surpassed the wild fish catch, but the cost of fish meal and oil to feed farmed fish and shrimp continues to rise as oceans are depleted. Spirulina and other algae supplemented feeds show better growth rates, improved quality and color, better survival rates, reduced medication and effluent waste.

Numerous studies have show spirulina benefits for farmed animals such as chickens, pigs and cows. Spirulina feed suppements raised anti-viral activity in chickens, reducing the need for antibiotics for healthier chickens. Adding a small percentage of spirulina to cow feed increased cow body weight, milk quantity and quality. Nannochlorospis algae is added to biofeed for chickens, pigs and cows for richer omega-3 fatty acid content in premium egg, milk and meat products.

***Recovering degraded cropland in Arizona with algae smartcultures. Comparing biofertilizer and conventional melons.** (Mark R. Edwards)*[12].

Algae biofertilizers restore soil fertility

A U.N. FAO report *Blue-Green Algae in Rice Production* revealed the possibilities of blue-green algae replacing chemical fertilizers and rebuilding the structure of depleted soils.[11]

In India, blue-green algae is grown in shallow earth ponds. When water evaporates, dried algae is scooped up and sold to rice farmers. This natural nitrogen source increased rice yield. Where chemical fertilizers are not used, algae gives the same benefit as 25 to 30 kg of nitrogen fertilizer per acre. Where chemicals are used, algae reduces chemicals by the same amount.

Algae rich in polysaccharides help recondition soil fertility and build structure to retain moisture. Algae have plant growth regulators, and by inoculating soil with algae, productivity is enhanced. Used as biofertilizers, algae can help feed people through soil renewal, providing economic opportunity without resorting to the vicious cycle of chemical fertilizers, soil exhaustion.

Farmers can buy algae biofertilizer treatments. In a field study in Arizona, live algae 'Smartcultures' flowing onto degraded land improved soil microbiome and grew superior melons. Vast areas of depleted and abandoned cropland can be regenerated quickly with algae biofertilizers, increasing land fertility and value.[12]

Food scanning tomatoes • QR Code • Scanning products

Ag technology, food scanning and food choices

Food production, distribution and sales are rapidly changing with ag technology and big data collection for precision agriculture, and tracking distribution. At the market, consumers can scan individual products for nutritional content and product information.

How will customers know nutrient dense algae-based products are a better buy than empty processed foods? DNA test kits and sensors will help consumers monitor their own microbiome. When scanning QR codes on products, their smartphone app can inform them which foods and products are best for their health. Advanced scanners will detect the molecular food signature and the app will translate that into nutritional content.[13]

Today, products have labels with QR type codes. Scientists are testing new kinds of biodegradable packaging using algae based films and algae inks. Articles now cite the potential of edible alginate, microalgae and spirulina micro coatings on fresh food, fruits and vegetables as natural bio preservatives retarding spoilage[14]. In the near future, these micro coatings could be imprinted with a QR code with algae based inks to be read by scanners.

Customers can quickly know where a tomato is from, how it was grown, its nutritional value and whether this tomato is best for their body's health. When people have the metrics to choose nutrient dense foods, food companies will be eager to provide them.

Microalgae to clean up agricultural wastes

Agricultural and animal waste streams represent costly pollution and environmental problems. Conventional methods for wastewater and sewage treatment are expensive and rely on high-cost chemicals and heavy inputs of energy. Around the world, municipalities, utilities and ag companies spend large sums to treat wastewater and sewage and remove pollutants and impurities.

Algae systems can be cost effective, sustainable, long-term solutions. Algae grow well off waste stream nutrients. Algae can be local species that naturally grow in that ecosystem. Algae transform nutrient-rich and oxygen-depleted water into oxygen-rich clean water and algae biomass. Just like nature recycles everything as food, microalgae can capture and recycle CO2 effluent, animal and plant wastes for biofertilizers, animal and aquaculture feed and valuable extracts.

Palm plantation and palm oil mill effluent.

One example is growing spirulina on palm oil mill effluent. Malaysia produces about 45% of the world's pal oil and effluent disposal is a massive challenge. Research indicated spirulina could be cultivated in a mixture of palm oil mill effluent with good productivity and pigment content and should be considered for scale up to commercial production.[15]

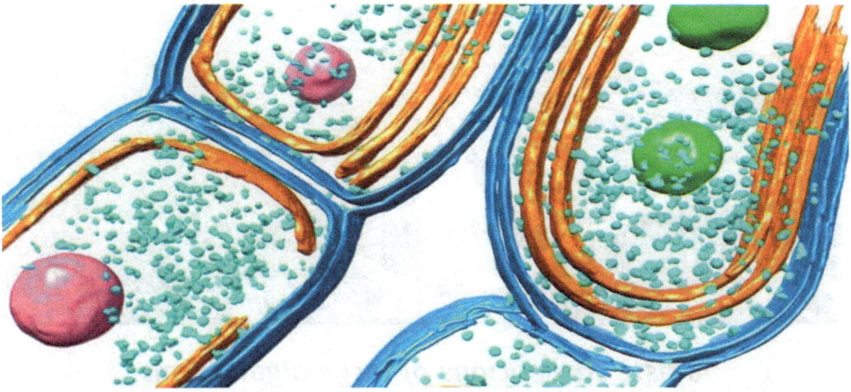

Illustration of cyanobacteria membrane. Tuomas Huokko et al.[16]

Cyanobacteria advances artificial photosynthesis

Plants have one major advantage over humans: They can make energy directly from the sun. If humans are able to mimic photosynthesis to harness the sun's energy for clean, storable, efficient fuel. If so, it could open a whole new frontier of clean energy.

The closest process to artificial photosynthesis today is photovoltaics, where a solar cell converts the sun's energy into electricity. That process is inefficient, capturing only about 20% of the sun's energy. Photosynthesis is way more effcient, capable of storing 60% of the sun's energy in biomolecules instead of batteries.

Scientists have discovered how cyanobacteria, or blue-green algae like spirulina, nature's first photosynthetic life form, performs photosynthesis in thylakoid cell membranes. This knowledge of how light is efficiently captured could be transferred to plant cell membranes to enhance photosynthetic performance, improving sustainable crop yields in an era of climate change.

Knowledge of cyanobacteria membrane pathways may also benefit the design and generation of artificial photosynthetic devices for efficient electron transfer and bioenergy production.

Artists conceptions of large algae farms.

Bio-solutions for climate change

Global warming is driving climate change. The world's most productive crop regions could fail to produce food because of excessive heat and lack of rainfall. Higher heat and drought has already been experienced across vast sections of Northern Africa and the Middle East to India. A long term drought cycle is feared for US Western states and the great North American grain belt.

These climate changes can trigger worldwide food shortages, loss of opportunity, political unrest, uncontrolled migration. This disruption is already underway and now there is a scramble to find solutions. Microalgae require less land and water than other protein foods and can grow in hot climates where other crops cannot.

While commercial farms grow spirulina as a health food for millions of people, microfarms in Africa, Asia and South America can produce food for local people. With appropriate technology, they address needs of waste treatment, soil and water quality, reforestation and food production, reviving environment and economy.

But these solutions are not at the scale needed to reduce the costs of algae protein to replace dependence on industrial agriculture meat and vegetable proteins. Another bio-solution is coming.

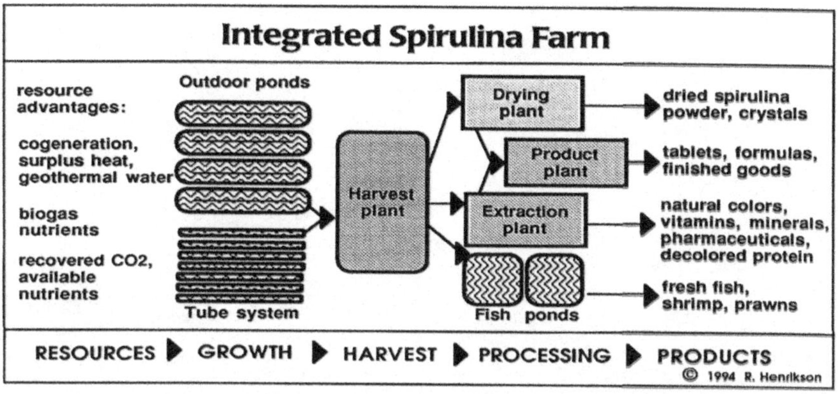

Integrated microalgae biorefineries

Future farms will integrate sources of nutrients and energy, and produce a variety of end products. Prototype plants are already underway, but none of these projects are currently at scale needed to dramatically reduce the costs of algae protein.

Biorefineries will use waste heat and CO2 from industrial factories, energy plants and data processing centers that power the cloud and cryptocurrency mining produce heat and have a huge carbon footprint. Recovered nutrients from waste streams from industrial and animal processing plants can feed algae systems.

Spirulina and other microalgae will be grown in both ponds and photobioreactors using both sunlight and solar photovoltaic light sources, depending on climate and location.

The harvest, drying, product and extraction plants will produce a wide variety of health food products, natural colors, alternative protein, and pharma products. Dried algae can be added to animal and aquaculture feeds. Fresh wet algae harvest can be used directly to feed adjacent fish growing and aquaculture systems.

At large scale, these integrated microalgae biorefineries will produce protein competitive with conventional proteins while reducing CO2 emissions and without degrading soil and water resources.

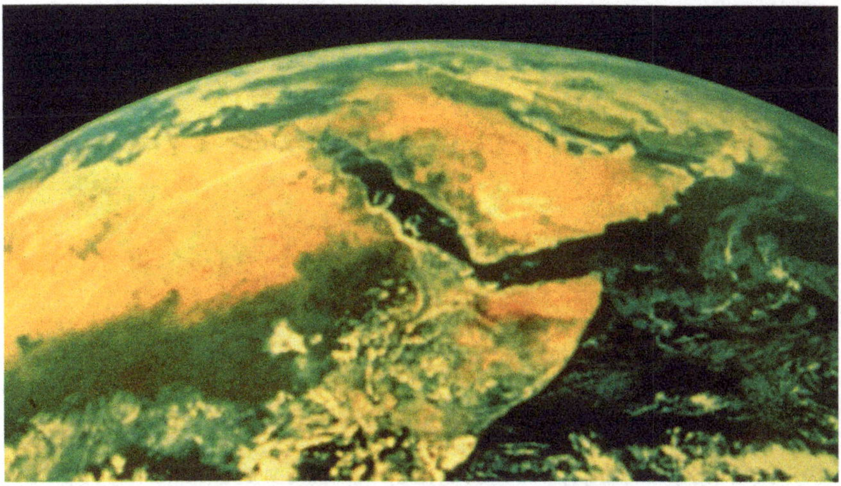

Chapter 9

Microalgae's role in restoring our planet

- **New algae products from the source of life**
- **Algae in space, on Mars, and back on Earth**
- **Farming alkaline lakes, regreening desert, sea farms**
- **Integrated vertical and urban farms**
- **Microfarm networks and ecological communities**
- **Beyond linear to circular production models**
- **Planetary restoration and individual transformation**

Understanding microalgae, the foundation of life, can help us develop restorative models of personal and planetary health. Algae is essential to Earth's self-regulating life support system.

We can imagine our algae future with schemes and dreams using microalgae to help regreen the desert, refertilize depleted soils, farm oceans and enhance biodiversity.

The big challenge is how to lower algae production costs: scaling up massive algae biorefineries within a circular bioeconomy.

Big investments in algae will grow our future food and its own bio-packaging

Over 40 years ago, the first entrepreneurs were building algae ponds iaround the world. The first algae foods were introduced into US stores in 1979. With 20 times the productivity as conventional crops, algae promised to be a *'food of the future'*.

The dominant growing technology for spirulina has been open ponds with paddlewheels in warm, sunny climates. Other microalgae are grown in photobioreactors or fermentation systems. But production costs have remained high due to a combination of factors: using agricultural land, fresh water, clean nutrients, skilled personnel, servicing big investments, harvesting and drying infrastructure, and complying with food and quality regulations.

Annual world microalgae output may well exceed 10,000 tons. But even big algae farms are relatively small, less than 100 hectares in size. Yet, the number and variety of food and specialty products from algae has flourished. Algae is an ingredient in thousands of products for food, feed, colors, nutraceuticals, medicinals, cosmetics and personal care, biofertilizers and fine chemicals.

Today's big investments in algae may take a decade to reach commercialization. We'll see healthy algae omega 3 oils and protein food and feed products, algae textiles, and algae based biopolymers and bioplastics replacing fossil fuel chemical products.

The Challenge: How will production costs come down?

Algae ventures have raised over a billion dollars for algae biofuel R&D and production. Innovations will dramatically change the way algae has been produced for 30 years. How?

• Discover better performing algae cultures. Scientists will identify algae with superior properties, faster growth rates, and grow in low light and temperature and high saline, brackish or ocean water.

• Develop simpler design and technology. Rethink, redesign and reengineer growing, harvesting, processing and drying to reduce capital costs, operating costs and power use.

• Use marginal land and water just like nature. To grow large scale, use remnant flat land and ocean, saline, brackish or waste water located near nutrient resources.

• Use waste nutrients just like nature. Capture and recycle CO2 effluent, animal and plant wastes. Ferment agricultural, animal, industrial and waste streams into carbon, nitrogen, phosphorus, potassium and trace nutrients to feed algae.

• Use all the algae biomass just like nature. Start with the end product and work backwards. What products can be sold, for how much, and what will be markets for those products?

• Create multiple revenue streams. Environmental services such as carbon credits, pollution mitigation, wastewater treatment, biomass and waste heat for electricity. Potential revenues include algae oil supplements for animal feedstocks, biofertilizers, fine chemicals and bioplastics, extracts for pigments and nutraceuticals.

• Scale up to thousands of hectares. Demonstration farms will prove out technologies will generate revenues for smaller markets.

• Exploit the unexpected- carpe diem. Investment drives innovation, creates serendipity and breakthroughs.

International Algae competition: algaecompetition.com.

International Algae Competition
A global challenge to design visionary algae systems

Imagine our future living in cities where buildings are covered with photosynthetic skins and vertical gardens, collecting the sun's energy and producing food and energy for urban citizens. Imagine greening desert coastlines and producing food for millions of people. Imagine algae systems that recycle polluting wastes into high value animal food, fuel and biofertilizers.

The International Algae Competition challenged the world : *"How will algae change the world and improve our lives".* Participants from 40 countries submitted entries of designs and actual projects for 1. Landscape Designs, 2. Production systems, 3. Food Development.

Imagine Our Algae Future Book @ amazon.com.

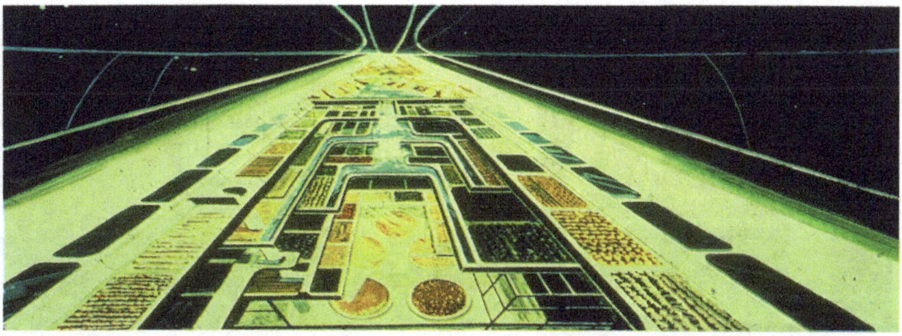

Artists vision of a space colony with fields and ponds.

Algae for life support systems in space

The International Space Station originally envisioned a Controlled Ecological Life Support System (CELSS), providing oxygen and food for humans and recycling wastes. Because designs left limited room for plants, scientists looked at algae.

Algae have higher photosynthetic efficiency, producing more oxygen and food than any plant. Nutrients come from carbon dioxide exhaled by humans and recycled wastes. This solves the problem of disposing of wastes during space travel. Algae turn waste into water, nutritious food and oxygen to support humans.

Ambitious plans included a system to grow algae to support humans, with lenses and optical filters to collect sunlight. The international Space Station was less ambitious. Instead of CELSS, rockets take supplies to the ISS and bring wastes back to Earth.

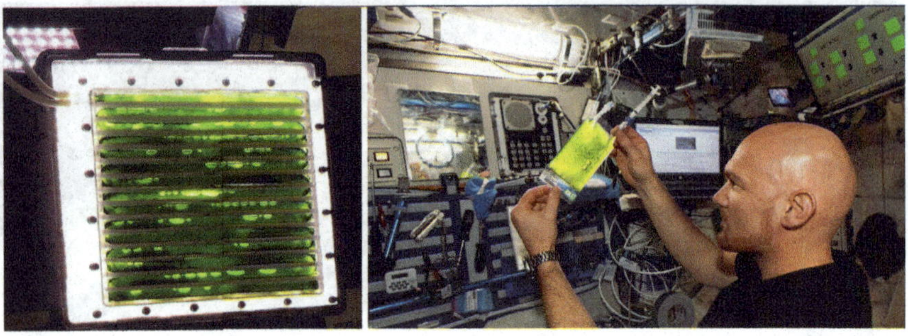

Algae photobioreactor[1] and algae experiments on the ISS.[2]

Algae on the International Space Station

As missions become longer and venture farther out into space, crews will have to grow their own food. Algae should survive well on space missions. Cold loving strains of green algae and blue green algae survived 16 months on the exterior of the international space station (ISS) despite extreme temperatures in the vacuum of space and UV and cosmic radiation.

The Melissa team in Europe has been working for the past decade using algae to recycle water and oxygen on the ISS to decrease the number of restock missions by experimenting with organisms like spirulina algae. An algae photobioreactor was sent to the ISS in 2018 to see how well the algae will grow and process carbon dioxide. It will join with the Advanced Closed-Loop System (ACLS) already on the ISS to produce breathable air.[3]

Spirulina will make a six-month trip to space in 2021 to test its suitability as a nutritious crop for astronauts on long-term space missions. Spirulina will grow in culture bags that allow diffusion of cabin air into the growth media. Astronauts will place the bags under a grow light. Every few weeks, the astronauts will transfer some of the algae into new media and take samples for analysis on Earth. The collected samples will allow researchers to test the algae to see if changes occur during extended periods of time in space.

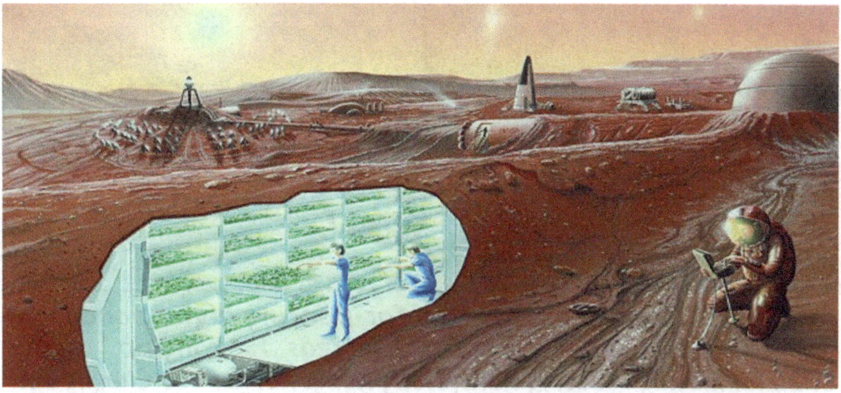

Biodomes growing algae and bacteria on Mars[4].

Growing algae on Mars for oxygen and food

Decades ago, scientists envisioned terraforming Mars so humans could colonize it. By adding nitrogen and oxygen to the atmosphere; pumping water to the surface; cooking for decades, spicing with cyanobacteria, then adding the rest of Earth's plants and animals. But terraforming might take 300 years or more.

More realistic goals by SpaceX and others is putting humans on Mars by 2030. A mission to Mars could require a three-year round trip. Algae will play a vital role in long-term space missions. Production of food on Mars would be essential for survival.

Cyanobacteria (blue-green algae) are crucial for life on Earth. 2.4 billion years ago cyanobacteria filled the atmosphere with oxygen, making subsequent life possible. Research has shown that cyanobacteria can use gases in the low oxygen and low pressure Martian atmosphere as their source of carbon and nitrogen.

Low-light adapted cyanobacteria can grow under rocks and potentially survive the harsh condition on Mars. One idea is using bacteria and algae with Martian soil as fuel to pump out oxygen. A Mars colony would have sheltered biodomes growing algae and bacteria and colonists could use this oxygen to survive.

Algae Competition designs for floating algae ponds.
Landscape for warm coastal areas. Ho Wing Ho.[5]
CO2 Eliminating Floating Park, Hong Kong. Adrian Yee Cheung Lo.[6]

Seawater and floating farms

Developing ocean water strains of spirulina would allow cultivation along arid and desert coastlines. There have been numerous designs and attempts with floating algae ponds.

In *Spirulina, Production & Potential,* Dr. Ripley Fox envisioned huge seawater farms along desert coastlines.[7] He proposed a network of 25 farms, 120 hectares each, providing 10 grams a day for 30 million children. Ponds lined with plastic film and floating paddlewheels circulating the ponds would be powered by solar ponds. CO2 would be recovered from power and industrial plants. He suggested farms be owned by NGOs and financed by governments.

Floating spirulina beds on pond in Malaysia.
Algae International Sdn Bhd.[8]

Flamingos feeding on spirulina in Lake Nakuru, Kenya.

Farming natural alkaline lakes

Spirulina lakes are found in Peru, Chile, Myanmar, Australia and stretch across the Sahara and East Africa, near millions of undernourished people. For 40 years, scientists have proposed harvesting algae from lakes in Ethiopia and Kenya. The best approach would be building cultivation ponds beside these lakes, without disturbing the larger ecosystem. Projects sponsored by governments and business could distribute spirulina to local people.

Farming African lakes could create opportunity: 1) help against AIDS, 2) export for hard currency, 3) source of a new food and 4) relieve pressure on food growing areas.

Greening desert coastlines

Large areas cannot be reforested if millions of people surrounding them cut down forests for food and fuel. Economic opportunity zones alongside reforestation zones will be critical to reforestation. Microalgae can create opportunity with a fraction of the land and water as conventional crops. Several restoration projects have been proposed. Once scientists learn to cultivate microalgae in seawater, new food growing areas can use 10,000 miles of accessible desert coastline in hot climates: Mexico, Peru, Chile, West, North and East Africa, Egypt and the Arabian peninsula and India.

Federation of Spiruliniers gathering.[9]

Spirulina microfarm networks

Hundreds of spirulina microfarms around the world in 30 countries produce fresh and innovative food products for their local region. Microfarms scale to any size and can be sited practically anywhere, including in cities.

Big commercial farms selling bulk spirulina powder to other companies may realize only 10% of the retail value of packaged products. Microfarmers who make their own branded products and sell locally direct to consumer can capture up to 100% retail value. Locally produced fresh and frozen spirulina sells for 2 to 3 times the value of dried spirulina.

More algaepreneurs are starting microfarms every year and forming networks to share information and resources. The largest established network is the Federation des Spiruliniers de France (FSF) with 160 members. FSF is a network of producers supporting each other and sharing the same values: mutual aid, respect for living things and the environment.[9]

FSF is an example of the spontaneous development of a new agriculture carried by small producers. Its objective is to professionalize microfarmers and maintain the link between its members. The FSF action plan is to improve knowledge of the product and its promotion through labels, provide support to producers through training and engineering research, encourage the consumption of spirulina within a healthy, sustainable diet.

Jourdan greenhouse. Making spirulina noodles. France 2002.[11]

Family and community cultivation

Many people have asked how to grow spirulina in their back yard. *Spirulina: Production & Potential,*[10] by Dr. Ripley Fox, reviews knowledge, equipment and funds needed, conditions for growth, harvesting and drying. The manual how to cultivate spirulina on a small scale is *Cultivez Votre Spiruline.*[11] For many years the author Jean-Paul Jourdan operated spirulina farms in Europe and Africa.

Fox and Jourdan catalyzed spirulina microfarmers in France starting a trend growing spirulina in greenhouses in developed countries with temperate climates. Today there are algaepreneurs offering videos and courses online how to grow spirulina at home in aquarium tanks or backyard ponds.

Ecological communities

Regenerative ecological systems need the efficiencies of algae at the base of the food chain, from the closed atmosphere of a space station to a developing world village. Ecological communities would be designed for high productivity, restoring the surrounding environment and will use smart technology with algae production, freeing up croplands for common areas or forests. A goal might be that for every acre farmed, another would be set free.[12]

Microfarms: Canary Islands, France, California, Mexico.

Peace Microfarms and Green Friendship Bridge

Peace Microfarms: A Green Algae Strategy to Prevent War[13], explains how algae microfarms give growers the freedom to produce food, feed and valuable bioproducts locally. Peace microfarms avoid war by producing food with minimal fossil resources, ending the need to fight over scarce food resources.

Peace microfarms distributed in food deserts across America could create food justice and give all children food security and health. Microfarmers could feed their families and communities.

Local microfarms can eliminate malnutrition and micronutrient deficiencies in rural and urban areas. A single 50 m2 microfarm can deliver enough Spirulina to cure 1,350 children and/or pregnant mothers from the curse of malnutrition.

The Green Friendship Bridge project describes how peace microfarms could be built across America, Mexico and Central America for only 2% of the cost of a proposed border wall with Mexico. The Green Friendship Bridge project examines the health, social, economic and environmental benefits of distributed microfarms.

Microfarms: India, Kenya, Chad, Togo.

Vertical farm greenhouses growing leafy greens and algae.

Vertical and urban farms

The growth of urban centers worldwide presents food security challenges from climate events, pandemics or economic disruption. Vertical farming, conceptualized in the 1990s, growing hydroponically in stacked layers with artificial lighting is now taking off. The dramatic impact of technology and cost reduction for lighting has led to massive investments in new vertical farms.

Because of high capital and operating costs, vertical farms need fast growing, high value crops like microgreens and herbs for fresh distribution to local stores, restaurants and consumers. Productive microalgae like spirulina would be a natural addition to vertical farms, improving economies of scale for operating costs and diversifying the product mix with fresh spirulina products.

Recently, Back of the Yards Algae Sciences in Chicago demonstrated spirulina phycocyanin extract worked as a biostimulant in hydroponics, improving yield, shelf life, color and taste quality of lettuce while cutting growth time by 21%.[14] A shortened growing period would improve sustainability of vertical farming by reducing energy and labor costs. It offers the possibility to extend indoor farming into more row crops as food staples ensuring better and more secure food supplies for urban populations.

2020 Urban Farm design for Bellingham Washington. [15]

2020 Net zero urban farm

This proposed organic urban aquaponics farm in Washington State on a remediated brownfield would produce fresh fish, vegetables, herbs and spirulina for the local market. Nutrient richwater from aquaculture tanks will irrigate and fertilize plants and then water is recycled back to the fish tanks in a closed loop system.

Self sufficient net zero building means it would use solar power and harvest using rain water for operations, netting zero draw from the energy grid and city water systems. This showcases an urban farm project on land not feasible for most other development.

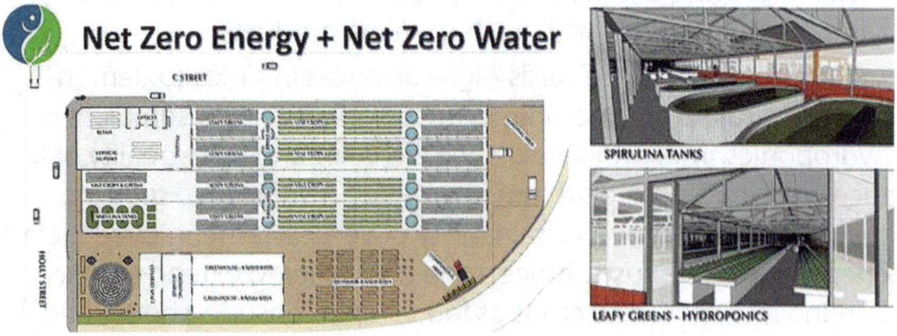

Layout of the urban farm showing spirulina growing system.

Algae greenhouse on top of Wincloud data center in Germany.[16]

Reducing carbon footprints with algae

Algae can nelp reduce carbon footprints. One example involves data centers around the world. They power cloud computing and even cryptocurrency mining, require a lot of energy and have a large carbon footprint. In data centres, the waste heat is blown out unused, which wastes around 95% of the energy used.

Data center operator Windcloud in Germany has developed with partners an innovative data center that works with electricity from wind power and uses waste heat for algae cultivation. The result is a neutral climate balance.

Exhaust waste heat had the ideal temperature for algae cultivation year round. A greenhouse was placed on top of the data centre, into which waste heat was pumped. In winter, algae can be grown with the high temperature of the exhaust air. But even in the summer, constant waste heat of 36 °C ensures optimal growth conditions for spirulina and chlorella algae. Harvested algae can also be dried with the exhaust air before the air is blown out.

After the successful commissioning of the data centre in August 2020, Windcloud is planning further data centres. This concept can be scaled for other locations worldwide to make a green change in the world of data centers.

Mother Jones Article 2016 (Clive Thompson).[17] • *Robocar.*

Regreening urban and suburban landscapes

This decade urban planners will have a second chance to reimagine urban design in a 21st century environment. Autonomous self-driving cars on demand, robot and drone delivery vehicles and taxi-bots will change life as we know it within 10 years. People won't own so many cars and park them while at work or shopping.

Today, parking takes up to 24% of the area of American cities. What will we do with those parking lots we don't need, when this real estate becomes available? Today we are at 'peak car'. How do we imagine our future cities without all those parked cars?

Maybe we will imagine a more human, greener eco-citiscape. Like abandoned railway tracks, parking areas can become green belts, restoring new life, energy and vitality to old cities. Urban farms can expand in agrihoods for more locally grown fresh food.

We can use those hard surfaces and cover them with vertical farms and greenhouse ponds for hydroponic, aquaponic and algae microfarm food production. Microalgae like spirulina are super-productive, and grow in very small areas for high-value products.

Joni Mitchell's "Big Yellow Taxi" came out in 1970 with the unforgettable line *"They paved paradise, and put up a parking lot".* Now, 50 years later, the line could be *"They made paradise, and greened up a parking lot".*

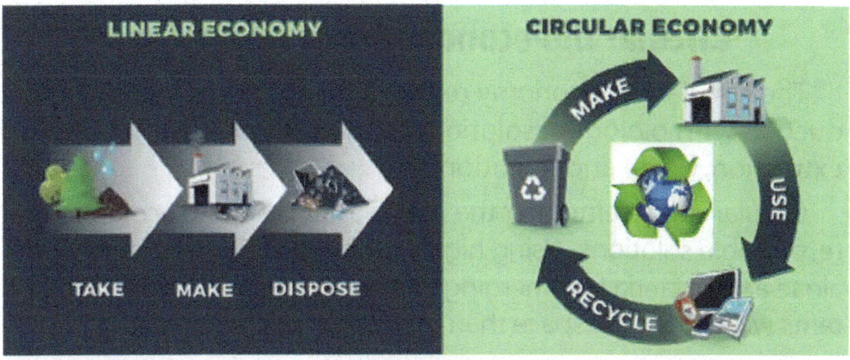

Linear and circular production models. © Peter Varga | Dreamstime.com

Beyond linear to circular production models

Agricultural and Industrial firms today follow an inefficient linear production model that relies on fossil resources that emit high levels of CO2 and levies a toll on human health and the environment.

Deforestation, clearing forests for cropland contributes about 20% of GHG annually to our atmosphere. GMO crops do not compete successfully with weeds and so farmers make wide use of herbicides, killing macro and micro biodiversity.

Intensive mechanical agriculture leaves waste to produce low energetic crops like maize where the waste biomass to food ratio is 97% waste, 3% food. Half the agri-inputs used to grow crops go to waste, including freshwater, fertilizer and pesticides. These wastes fill waterways with chemicals and pesticides.[18]

The linear economy model illustrates the waste caused by industrial production. Modern agriculture takes fertile land, fresh water, fossil fuels, inorganic fertilizers and pesticides and pollutes and degrades land, air, rivers, lakes and oceans.

A circular economy models nature. Natural resources are neither extracted nor wasted. Nature wastes nothing. Every living organism lives a life. At the end of life, it provides the nutrients for the next cycle of life.

Circular bio-economy with microalgae

The circular bio-economy replaces mechanical industrial production with biological solutions that avoid natural resource extraction, waste and pollution.

Circular biosystems offer the most sustainable and ecologically responsible solutions. Using highly productive and efficient microalgae and bacteria for emerging plant based food products, biosystems will use far less space than conventional agriculture.

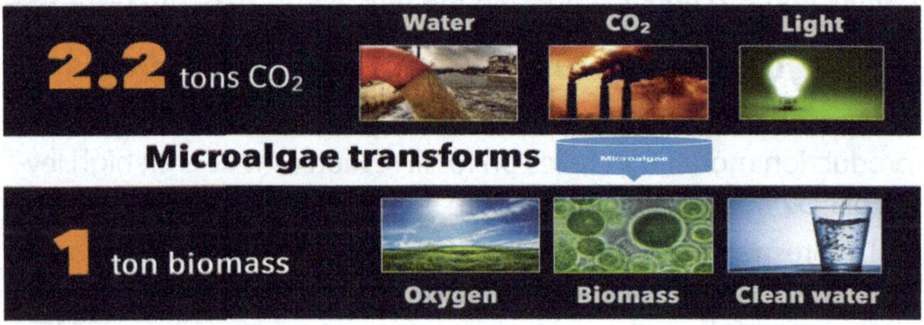

Microalgae transforms CO2 to biomass (David Punchard).

The first prototype sustainable microalgae biorefineries are now being funded. With microalgae, growers use light and photosynthesis as the engine instead of fossil fuels. Sunlight may be augmented with photovoltaic and concentrated solar.

At scale, microalgae biorefineries would capture CO2 as the primary nutrient from industrial sources or directly from the air. Other mineral nutrients would be recovered from waste streams. Algae produce oxygen, biomass and clean water. The algae biomass would be fractionated for protein, bioactive compounds and hundreds of byproducts.

By growing our food and many other products with microalgae, using less than 10% of the land area than conventional crops, this releases all that land for rewilding, new forests and carbon capture.

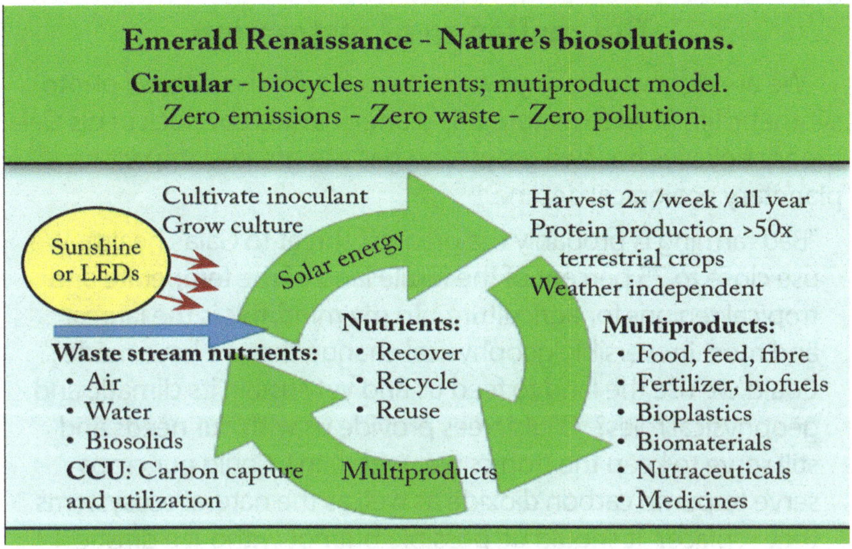

Microalgae in a circular bioeconomy (Mark R. Edwards).[19]

Full resource accounting for a new bioeconomy

Future integrated algae biorefineries need to operate like nature or as they would on a long term space mission. Everything has to be reused and turned into oxygen, clean water and food. This is the foundation of the circular bioeconomy we need for our planet.

Technologists are developing the metrics to understand the true costs of food production, including costs of CO2 emission and climate change, chemical pollution, lost fertile land and fresh water, human disease from nutritionally empty foods, and full product life cycle analysis. When these costs are applied this will make a level playing field for real market prices of food.

Compared with the linear model of intensive mechanical agriculture, a circular bioeconomy based on the advantages of microalgae production will produce abundant food and a variety of products at a lower cost while benefitting producers, people and planet.

The challenge of restoration

We are discovering ways we can work with the original photosynthetic life form to restore this planet. James Lovelock in his Gaia theory believes how we grow food has the greatest impact on planetary ecological decline.

> "Bad farming is probably the greatest threat to Gaia's health. We use close to 75 percent of the fertile land of the temperate and tropical regions for agriculture. To my mind, this is the largest and most irreversible geophysical change that we have made ... Could we use the land to feed us and yet sustain its climatic and geophysical roles? Could trees provide us with our needs and still serve to keep the tropics wet with rain? Could our crops serve to pump carbon dioxide as well as the natural ecosystems they replace? It should be possible but not without a drastic change of heart and habits."[20]

More people realize they can affect global food patterns by changing their own habits. Eating lower on the food chain, less meat, more organic vegetables and grains and even algae. They are healthier, help reduce environmental damage, and can help return cropland and grazing land back to new forests.

Choices are making a difference already

Consumer purchasing decisions make a difference. People will pay more for ecological food. Green business is good business. A company's socially responsible practices are one of the factors in deciding whether or not to buy. More of us understand Earth's biosphere is the foundation of our society and economy. People understand the connection between personal health and the planet's health, and they are making choices to help restore our health.

Microalgae like spirulina, are essential for individual and planetary health and restoration. The oldest photosynthetic life form is back - a return to the origins of life.

Procession

Spirulina speaks to the human species, on behalf of the first species – algae.

"We, the oldest algae, began photosynthesis at our greatest moment in evolution to initiate the unfolding of life. We can assist you, for we have a fundamental role to play.

Sometime in the future, you will look back at these years. The transformation of your attitude toward your body, your species and your planet will seem to have been remarkably rapid. This wonderful adventure will be the most magnificent and creative moment of your evolution in becoming fully human.

Understanding your evolution is so simple yet so profound. In realizing what harm you have done to our planet will come liberation from the past and your greatest gift to life. You will understand your role is not to control, but to participate with nature.

We are all composed of the same elementary particles dating from the beginning of creation. Your human form is yet another alignment of particles which dance together with all particles. Rejoice in your connection with all life.

In rediscovering your relationship with the natural world, you will heal yourself and our planet. We are excited for your opportunity in the coming years. Now go forth and express yourselves and participate fully in the beauty, the wonder and the glory of our unfolding creation." - ***Spirulina, circa 1989.***

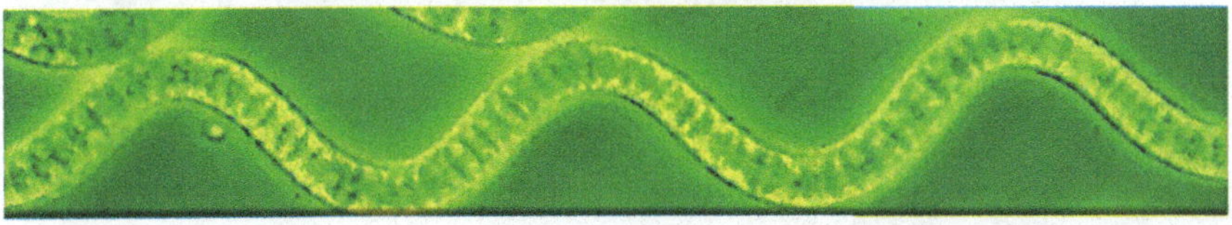

Appendix A
Quality and safety standards

Spirulina has been marketed and consumed as a human food and approved as a food for human consumption by governments, agencies and associations of these countries:

Argentina	Australia	Austria
Bahrain	Bahamas	Bangladesh
Belarus	Belgium	Brazil
Bulgaria	Canada	Chad
Chile	China	Colombia
Costa Rica	Croatia	Czech Republic
Denmark	Ecuador	Egypt
Ethiopia	Finland	France
Germany	Greece	Guam
Gulf States	Haiti	Hong Kong
Hungary	India	Iceland
Indonesia	Ireland	Israel
Italy	Jamaica	Japan
Kenya	Korea	Kuwait
Liechtenstein	Luxembourg	Macedonia
Malaysia	Mexico	Myanmar
Monaco	Netherlands	New Zealand
Nigeria	Norway	Peru
Philippines	Poland	Portugal
Romania	Russia	Saudi Arabia
Singapore	Slovenia	South Africa
Spain	Sweden	Switzerland
Taiwan	Thailand	Togo
Turkey	Ukraine	United Kingdom
United States	Venezuela	Vietnam
Yugoslavia	Zaire	Zimbabwe

Food safety research

Spirulina has a history of use in Chad where people consume 9-13 grams per meal, and these meals are from 10 to 60% of the meals.[1] "The U.N. FAO was attracted by the fact that algae was consumed by humans. The FAO organized an educational campaign in Chad to encourage consumption of spirulina harvested from natural sources. More than 6000 meals were distributed under the FAO and the campaign was a success, but was suspended due to the outbreak of war."[2]

Spirulina food safety guidelines
used by US and Japan growers

Criteria	*USA*[a]	*Japan*[b]
Moisture	< 7 %	< 7 %
Bacteriological		
Standard plate count	<200,000/ g	<200,000/ g
Mold	< 100 / g	< 100 / g
Yeast	< 40 / g	< 40 / g
Coliforms	neg.	neg.
Salmonella	neg.	neg.
Staphylococcus	neg.	neg.
Heavy Metals		
Lead	< 1.0 ppm	< 1.0 ppm
Arsenic	< 1.0 ppm	< 1.0 ppm
Cadmium	< 0.05 ppm	< 0.05 ppm
Mercury	< 0.05 ppm	< 0.05 ppm
Insect fragments	< 30 / 10 g[c]	*
pesticides	neg.	neg.
herbicides	neg.	neg.
additives	neg.	neg.
preservatives	neg.	neg.
dyes	neg.	neg.
stabilizers	neg.	neg.
artificial ingredients	neg.	neg.
fillers	neg.	neg.

a. Published by Earthrise Farms, 1995.
b. Published by Dainippon Ink & Chemicals, Inc., Japan.
c. U.S. Food and Drug Administration guideline.
definitions: < = less than, * = no set standard, / g = per gram, neg. = negative

Another report stated "*dihé* (spirulina sauce) was served at the school canteen. The introduction in the young people's food gave no problem in this region where the majority were Kanembou. But equally at Fort Lamy (now Ndjemena) the product was accepted.

Spirulina was given to malnourished children and adults in clinical studies beginning in the 1970s. Since then, millions of people in the developed countries have used it as a health food supplement, taking 3 to 20 grams a day. Rarely are there any reports of allergies.

In the 1970s, spirulina passed safety studies with animals and fish. Tests in France, Mexico and Japan showed no undesirable results and no toxic side effects on humans, rats, pigs, chickens, fish and oysters. Many rat feeding trials were held in Japan and no negative effects were found for acute or chronic toxicity or reproduction.[4,5]

In 1980, important animal studies were sponsored by the U.N. Industrial Development Organization (UNIDO) on rats and mice. Spirulina comprised 10% to 35% of the diet. No second or third generation reproduction, fertility, lactation, birth defect or cancer causing properties were found. No problems with heavy metals, nucleic acids, pesticides or bacteria. The study concluded any further research would show complete safety as a human food.[6]

Toxicology research continued in the 1980s and 1990s, showing spirulina has no toxicity in rats, no adverse effects on reproduction, including male and female fertility, duration of gestation, and no increase in abnormal offspring.[7,8,9]

Spirulina food standards
microbiological quality requirements of France, Sweden, Japan and Earthrise Farms (USA)

Standard	France[a]	Sweden[b]	Japan[c]	E Farms[d]
Moisture	*	*	<7 %	<7 %
Standard Plate Count	<100,000/g	1,000,000/g	<200,000/g	<200,000/g
Mold	*	<1000/g	*	<100/g
Yeast	*	*	*	< 40/g
Coliform	<10/g	<100/g	neg.	neg.
Salmonella	neg.	neg.	*	neg.
Staph	<100/g	<100/g	*	neg.

a. Superior Public Hygiene Council of France, 1984, 1986.
b. Ministry of Health, Sweden.
c. Japan Health Foods Association.
d. Earthrise Farms, 1995.

definitions:
< = less than
/g = per gram
* = no set standard
neg. = negative

Nucleic acid safety research

Spirulina has 4% nucleic acids (DNA and RNA), lower than chlorella and other microalgae, yeast and fungi (6-11%). Although there was once some concern microalgae might increase uric acid levels because of nucleic acids, there is little evidence to support this. One study found uric acid levels did not increase in humans taking up to 30 grams a day of 50 grams of chlorella.[10] Since spirulina is lower in nucleic acid content, eating up to 50 grams a day is safe as well, and means it can be safely used as major protein source.[11]

Independent labs around the world confirm the absence of any toxic effects even when it provides a significant amount of dietary protein.[12,13,14] Since its introduction as a human food in 1979, Spirulina has been safely consumed by millions of people in North and South America, Asia, Europe and Africa.

Heavy metal safety research

Mercury, lead, cadmium and arsenic are widespread in our environment from industrial pollution. Heavy metals are toxic in small amounts, and prolonged eating of foods with heavy metals can lead to long term health problems. Yet few companies disclose levels of heavy metals in foods. Earthrise Farms published standards for heavy metals in spirulina.[15] Five year testing in California showed heavy metals were not detectable or extremely low. Based on 120 independent lab tests, Earthrise set some of the toughest standards for heavy metals.

Mercury was not detectable in 40 tests, and the standard for mercury was set at less than 0.05 parts per million (ppm). In comparison, the US FDA standard in 'aquatic animals' is 1.0 ppm, permitting over 20 times more mercury. Standards were set for cadmium (less than 0.05 ppm), lead (less than 1.0 ppm), and arsenic (less than 1.0 ppm). By comparison, the UN Protein Advisory Group standard for single cell protein permits higher heavy metals: 1.0 ppm mercury; 1.0 ppm cadmium, 5.0 ppm lead; and 2.0 ppm arsenic.

Heavy Metal Guidelines

	Lead	Mercury	Cadmium	Arsenic
	< (less than) parts per million (ppm)			
Aquatic Animals				
(U.S. Food and Drug Admin.)	-	<1.0	-	-
Single Cell Protein				
(UN Protein Advisory Group)	<5.0	<0.1	<1.0	<2.0
Chlorella				
(Japan Health Foods Assn)	-total heavy metals: < 20.0 ppm			<2.0
Spirulina				
(Japan Health Foods Assn)	-total heavy metals: < 20.0 ppm			<2.0
Spirulina				
(Earthrise Farms guidelines)	total heavy metals: < 2.1 ppm			
(*undetectable)	<1.0	*<0.05	*<0.05	<1.0

Tolerable amount per day

	Lead	Mercury	Cadmium	Arsenic
	mcg (micrograms) per day			
UN World Health Organization				
/ Food & Agriculture Org.	500	50	66-83	-
Earthrise Spirulina (10 gm per day)				
mcg of heavy metals:	<10	<0.5	<0.5	<10
percent of UN Guidelines:	<2%	<1%	<1%	-

Algal toxin safety research

A quality control issue for blue-green algae (cyanobacteria) is inadvertantly harvesting other blue-green algae containing cyanotoxins. This is a risk when harvesting algae from natural bodies of water with mixed cultures of microscopic algae. Algal toxins are can cause widespread poisoning of animals and humans.[16]

In 1995-96, microalgae producers sponsored research conducted by algal toxicologists. The result was a *Technical Booklet for the Microalgae Biomass Industry* as a guide to a enzyme linked immunosorbant assay (ELISA) and a protein phosphate inhibition assay (PPIA) for detection of toxic microcystins. These methods detect, monitor and control cyanotoxins to assure a safe product.[17]

Quality and Safety Standards for Spirulina for the USA Natural Foods Industry

The Natural Products Quality Assurance Alliance (NPQAA) and The Natural Nutritional Foods Association (NNFA)

Definition (dried powder as produced)

Spirulina is a blue-green microalga. Spirulina is cultivated in specially designed artificial ponds, harvested, dried and packaged in accordance with Good Manufacturing Practices (GMP), and quality controlled at each stage of the production process.

Legal Definition

According to the U.S. FDA (Talk Paper 6/23/82): as a food, spirulina can be legally marketed as long as it is labeled accurately and contains no contaminated or adulterated substances.

Spirulina powder as produced

Physical Appearance and Identity for spray-dried spirulina.
Spirulina should be readily identified as *Spirulina (Arthrospira) sp.* under microscopic and biochemical examination. Spirulina powder is fine uniform powder, dark green in color, with mild seaweed taste and with no decayed or bitter taste or smell.

Minimum Nutritional Content

Component	Amount	Analysis Method
1. Protein	55 %	AOAC
2. Total Carotenoids	300 mg/100g	AOAC modified
3. Chlorophyll-a	900 mg/100g	AOAC modified
4. Phycocyanin	8,000 mg/100g	DIC method
5. Vitamin B-12	200 mcg/100g	AOAC (microbio. assay)
6. gamma-linolenic acid	900 mg /100g	AOAC
Moisture. Acceptance criteria for each production lot:		
1. Moisture	less than 7%	AOAC
Bacteriological Assays. Acceptance criteria for each lot:		
1. Standard Plate Count	less than 200,000/g	FDA Bacteriological Manual
2. Molds	less than 100/g	FDA Bacteriological Manual
3. Yeast	less than 40/g	FDA Bacteriological Manual
4. Coliforms	less than 3/g	FDA Bacteriological Manual
5. Salmonella	negative	FDA Bacteriological Manual
6. Staphylococcus	negative	FDA Bacteriological Manual

Quality and Safety Standards for Spirulina for the USA Natural Foods Industry

The Natural Products Quality Assurance Alliance (NPQAA) and The Natural Nutritional Foods Association (NNFA)

Extraneous Materials.

For USA human consumption only, testing of each production lot is required. *US FDA Guideline acceptance criteria.

1. Insect fragments	*less than150/50g	AOAC (1990) 15th ed.
2. Rodent hairs	*1.0/100g	AOAC 990.09

Heavy Metals. Shown by a typical analysis of spirulina:

1. Lead	less than 2.5 ppm	AOAC
2. Arsenic	less than 1.0 ppm	AOAC
3. Cadmium	less than 0.5 ppm	AOAC
4. Mercury	less than 0.05 ppm	AOAC

Supplementary Guidelines. Shown by a typical analysis of spirulina:

1. No pesticides	4. No preservatives
2. No herbicides	5. No stabilizers
3. No dyes	6. No irradiation

Spirulina finished products

Finished products for human consumption shall meet all relevant USA food quality and safety standards, and shall follow the appropriate Good Manufacturing Practice Guidelines.

Minimum Nutritional Content. To be determined.

Moisture. Acceptance criteria for each production lot:

Moisture	less than 7%	AOAC

Bacteriological Assays. Acceptance Criteria:

1. Standard Plate Count	less than 200,000/g	FDA Bacteriological Manual
2. Molds	less than 100/g	FDA Bacteriological Manual
3. Yeast	less than 40/g	FDA Bacteriological Manual
4. Coliforms	less than 3/g	FDA Bacteriological Manual
5. Salmonella	negative	FDA Bacteriological Manual
6. Staphylococcus	negative	FDA Bacteriological Manual

Product shelf life:

Producers of finished products shall determine nutrient statements on labels based on both bulk spirulina powder analysis and nutrient changes due to tableting and bottling and package shelf life.

Appendix B

Personal journey from small to large to small again

Origins of Earthrise - Early spirulina farming in the USA

Larry Switzer, visionary bioneer and catalyst, founded Proteus Corporation in 1976 to develop spirulina algae as a world food resource. Proteus was funded by private California investors, committed to the spirulina vision.

Larry Switzer and the little spiral.
Spirulina article by Eric Perlman, San Francisco Examiner, 1977.

Hope for the world's children

Switzer was looking for solutions. He discovered microalgae was 20 times more productive as a protein source than any other food and could be grown on unused land and water. It was possible to cultivate on a large scale. Scientists discovered spirulina was safe, had been consumed for hundreds of years by traditional peoples, and showed promising nutritional health benefits. If this algae were consumed by millions of people, it would have tremendous benefits especially for the world's children and our planet's future.

However, no one had cultivated spirulina on a large scale, produced it as a food, and convinced anyone they should eat algae! If it was an idea whose time had come, it was a daunting task.

Larry was joined by Robert Henrikson and Dr. Alan Jassby, who helped design the early systems and cultivation programs. In early 1977 Larry and Robert left Berkeley for California's hot Imperial Valley, ideal for growing spirulina. For three years this small team sweated in the desert to build a successful farm model. Just about everything that could go wrong, did, and every problem had to be overcome.

Farm site 1977. Growing up the culture in trays and pools.

First prototype farm in the California desert 1977

The first paddlewheel growing ponds were underway in May, but then, in August 1977, our first farm was wiped out by Hurricane Doreen! This was a one-in-a-hundred year storm and our little farm was in the middle of 10 square mile flood under 4 feet of water.

Robert with the first paddlewheel pond. The Flood of 1977: On a raft and diving under water to retrieve pumps.

Building the second prototype farm

We asked our investors for more funding, relocated, and started a second spirulina farm. Ron Henson showed up to help with construction, and eventually became Ops Manager of Earthrise Farms and Sales Manager of Earthrise Nutritionals.

Second pilot farm. 1979.
Robert Henrikson., Ron Henson, Bruce Carlson, Larry Switzer.

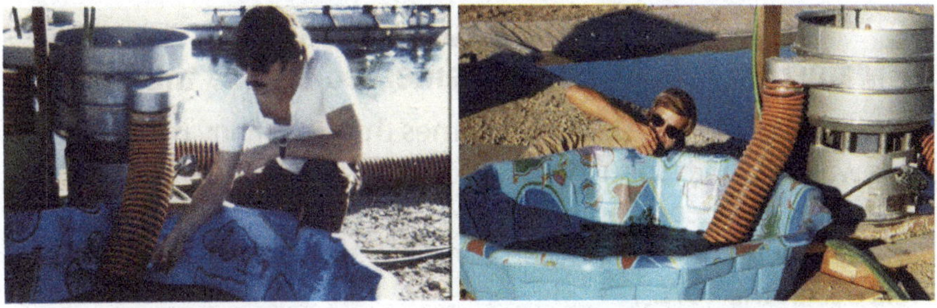

Larry inspects and Robert eats spirulina fresh harvest.

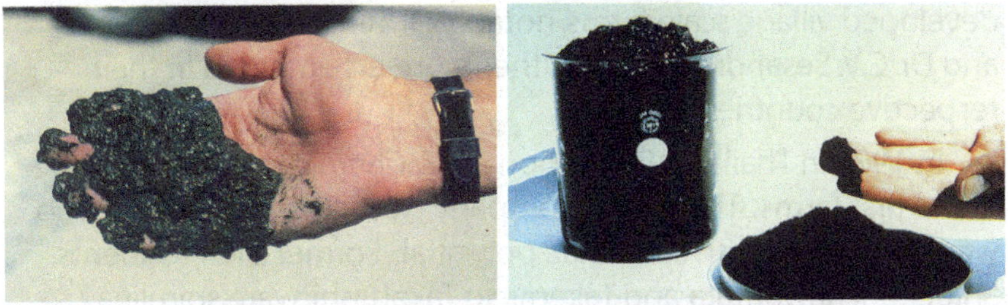

Fresh harvested paste has a mild taste.
Fresh paste and spray dried powder.

Paddlewheels circulate the water around raceway ponds.

We developed larger growing ponds, tested harvesting and drying operations. Then we began looking for a new funding source for a commercial farm, several million dollars. To interest investors, we had to prove we could sell algae. We began importing spirulina, and developed a partnership with a Japanese company that had just begun growing it in Thailand.

Fortunately we were not alone. Many other algae bioneers emerged around the world.

A new planetary idea often comes through many messengers. This was true with spirulina. While Earthrise was underway in California, other companies began cultivation. Hubert Durand-Chastel encouraged a Mexican company to set up a farm in Lake Texcoco in the 1970s.

Israeli, Indian and European scientists began research. Others developed village scale farms, notably Dr. Ripley D. Fox of France, and Dr. C.V. Sesahdri of India. Other bioneers emerged in their respective countries.

In 1980 in Thailand, Dainippon Ink & Chemicals (DIC), built one of the first farms. This global company was led by visionaries who were fascinated with spirulina's potential. Former DIC Presidents, Shigekuni Kawamura and Takemitsu Takahashi, were spirulina sponsors, and funded development. Heading up the program was "Mr. Spirulina" in Japan, Hidenori Shimamatsu of DIC Bio Division.

Earthrise® spirulina at natural food trade shows 1979.

Introducing algae to the natural food market

In 1978, we chose Earthrise as our trademarkl- the rising of the Earth from the moon's surface in 1969. This image represents our awakening. We rediscovered blue-green algae, offering health benefits to ourselves, our society and our planet. We dedicated these gifts to a new Earth Rising.

Earthrise sales were directed by Robert Bellows and Terry Cohen of Boulder, Colorado. Earthrise® began in natural food stores in 1979 and gained popularity fast. In 1981, the National Enquirer touted it as a magic diet pill, and consumer demand exploded. Diet companies jumped on the bandwagon, and sold spirulina pills that didn't even contain any. Not much real spirulina being grown.

The diet boom faded but by 1987 the market began to grow again. More people experienced health benefits. Published scientific research documented its therapeutic benefits.

Earthrise® Teams.

Earthrise Farms expansion over the years.

First U.S. production: Earthrise Farms 1982

We built a relationship with DIC as we imported from their farm in Thailand for Earthrise. This partnership between California entrepreneurs and Japanese corporate intrapreneurs blossomed. We shared a common vision of spirulina's coming impact on the world. Together we founded Earthrise Farms, the first US production farm, as a joint venture in 1982.

Earthrise Farms became a subsidiary of DIC. Under the leadership of Yoshimichi Ota, the farm expanded. Dr. Amha Belay from Ethiopia, became VP, responsible for cultivation and Juan Chavez, Production Manager, raised output to 500 tons a year by 1996.

Earthrise Farms 1984: Bellows, Henrikson, Ota, Hamada, Carlson, Shimamatsu, Jassby, Sonneville. 1996: Chavez, Ota, Belay.

Earthrise Spirulina ad evolution: 1981, 1990, 1997.

Earthrise products go global 1990s

In the 1990s, Earthrise Company grew an extensive line of spirulina food supplements for US natural food stores. By 1996, Earthrise® trademark products were sold in 40 countries, making it the world's best selling spirulina at that time.

Earthrise developed an extensive line of food supplements for the US market in about 10,000 natural and health food stores, mail order and other outlets.

Earthrise product line 1990s.

Costa Rica • Taiwan • Canada • Slovenia

Germany • China

Aquaculture feed • Dog and Cat • Birds

Ripley and Denise Fox and assistants. Togo 1988.

Visiting a village farm in Togo 1988

Earthrise sponsored algae humanitarian projects in the 1980s and 90s. One was the integrated health and energy system in a village in Northern Togo. Dr. Ripley Fox developed an experimental system using appropriate technology to convert village waste through biogas digestion into nutrients for growing spirulina. Visiting the Togo project in 1988, I learned how small algae farms could play an important role in benefiting a local community.

People of Farende Togo. Collage by R. Henrikson.

öS 28,-
SFR 4,-

bulletin

Der Informationsdienst für gesundheitsbewußte Menschen 11/97

Erfolge mit Mikroalgen in der Naturheilpraxis

Spirulina in der Therapie

VON SABINE GERICKE

Spirulina ist eine winzige, spiralförmige Alge, reich an Proteinen, essentiellen Aminosäuren, Vitaminen, Mineralstoffen und essentiellen Fettsäuren wie der Gamma-Linolensäure. Sie wird kommerziell hergestellt und auf der ganzen Welt als Nahrungsergänzung im Gesundheitshandel vertrieben. Bis vor Senkung des Cholesterinspiegels und Einflüssen auf das Krebsgeschehen bis hin zur Stimulation des Immunsystems, Vermehrung der Lactobazilli im Darm, verminderter Toxizität von Schwermetallen und Medikamenten in den Nieren, bis zum Schutz vor Strahlung reicht.

Children of Chernobyl 1995

Earthrise and DIC donated spirulina to hospitals in Ukraine and Belarus. Doctors reported giving children 5 grams of spirulina per day reduced urine radioactivity by 50% in 20 days and concluded spirulina decreased radiation from contaminated food and was advisable to treat people with radiation sickness. In 1995, I visited hospitals and met with doctors and children in Ukraine, imagining how much better it would be if this region had its own algae farms.

At the children's hospital in Kiev Ukraine.

Philippe Calamand, Ripley Fox, Denis von der Weid view La Capitelle microfarm, 2002.

First spirulina microfarm in France 2002

At a conference in Mialet in the South of France in 2002, Jean-Paul Jourdan demonstrated growing spirulina in his personal greenhouse on his farmstead. Then the group moved to the new La Capitelle spirulina microfarm, started in 2001 by Philippe and Estelle Calamand. This was the first spirulina microfarm in France. Philippe showed how he grew, harvested and processed spirulina on a small scale making his own products that he sold directly in his local community. Within a decade there were over 100 spirulina microfarmers in France.

Mialet 2002. Ripley Fox, R. Henrikson, Jean-Paul Jourdan, Philippe Calamand, Hubert Durand-Chastel, Hendrik van Poederooijen.

Algae Tourism in Thailand 2010

Boonsom Spirulina Farm near Chiang Mai is the largest in Thailand. The founders are Professors Jiamjit and Somchye Boonsom. Beyond Green Diamond natural food supplements, the farm has an inviting health spa, offering spirulina waffles, ice cream, beer and facial masks. Boonsom Farm represents an excellent business model showing the benefits of algae tourism.

Lake and Pond Harvest in Myanmar 2013

One of the world's largest producers from both production ponds and natural lakes is in Myanmar. Started by Dr. Min Thein, June Pharmaceuticals harvests about 200 tons of spirulina per year and sells supplements, functional foods and beverages, cosmetics and personal care products, biofertilizer and spirulina beer.

Tour of Microfarms in France 2011

By 2011, 100 small spirulina growers stretched from Southern France to Normandy. I visited some of these microfarmers. Met with members of the Spirulina Federation and learned of their progress, challenges and prospects. Then in Normandy visited Eco-Domaine, an integrated experimental farm with a biogas plant.

Laurent Lecesve at Eco-Domaine microfarm in Normandy.

Lecesve, Planchon, Jourdan, Henrikson with the Eco Domaine team in Normandy. Ripley and Denise Fox hosted members of the Fédération des Spiruliniers in Laroque.

Imagine Our Algae Future 2012 • Algae Microfarms 2013

Imagine Our Algae Future 2012

The 2011 International Algae Competition hosted by Robert Henrikson and Mark R Edwards PhD was a global challenge to design visionary algae food and energy systems. *'How will algae change the world and improve our lives?'* Participants represented 40 countries and many of the best entries were featured in the book *Imagine Our Algae Future* published in 2012.

Algae Microfarms 2013

Microfarms around the world were featured in the book *Algae Microfarms* published in 2013. Microfarms for home, school, community and urban gardens, rooftop, mobile and vertical farms and living buildings. How algae microfarms can help transform our food culture by growing abundant healthy food in a very small area, extend the growing season, affordably and profitably.

The author's home growing systems: indoor tanks 2012, outdoor deck 2013, rooftop 2019.

Olympia Microfarm 2013-15

One of the northern most spirulina farms showed the ability of growing algae in cooler climates. Ponds had insulating foam panels below and retractable cover above, inside a vegetable greenhouse to help keep algae water warm, increase productivity and extend the growing season, beyond summer months. Monitored remotely by webcam with daily data sent from the local operator.

Northern California Microfarm 2016-19

This first NorCal spirulina farm in Half Moon Bay, south of San Francisco, operated inside a large commercial glass greenhouse once used for growing flowers. Spirusource fresh, frozen and dry products are sold in natural food stores in the SF Bay Area, direct to customers locally and online.

Spirulina Word Food International Editions

In 1989, Robert Henrikson published *Earth Food Spirulina.* Translated and published in Spain ('94), China ('95), Slovenia ('96), Macedonia ('97), Italy ('98), Taiwan ('99). 7th edition *Spirulina World Food* in 2010. Myanmar edition ('12). 8th edition 2021.

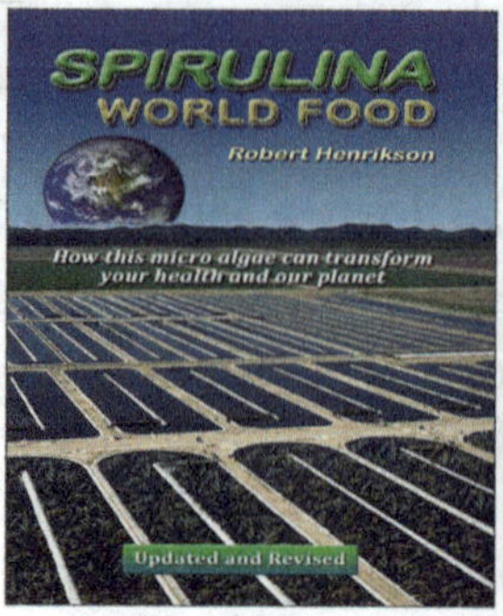

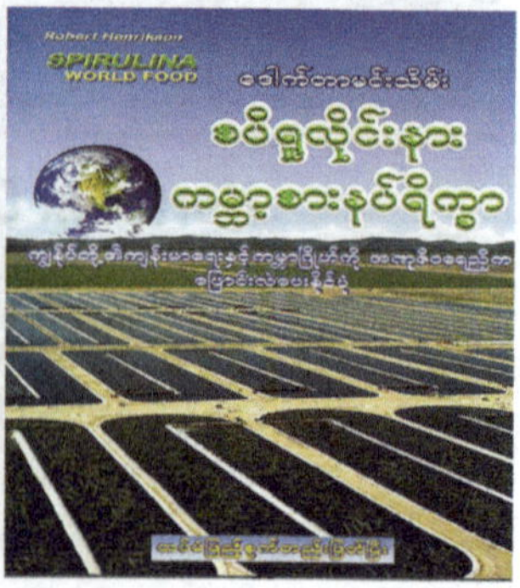

Spirulina Algae Video Documentaries

Over 40 years Robert Henrikson has produced short videos documenting the evolution of spirulina, from the first commercial farm to farms around the world today. Videos about health benefits of spirulina, conversations with algae bioneers, algae microfarms, village farms and visions of our future with algae.

See videos at:

spirulinasource.com • smartmicrofarms.com • algaecompetition.com

Robert Henrikson

For 40 years Robert has been an entrepreneur in algae, bamboo, carbon and natural resources. A pioneer in algae production and marketing, he was founder and director of Earthrise Farms, world's largest spirulina algae farm. For 20 years as President of Earthrise Company, he promoted Earthrise® Spirulina and superfood products in 30 countries.

Robert has visited algae farms and companies all over the world. He advises algae entrepreneurs, companies, investors and law firms in algae production, business plans and development, and marketing strategies. Founded International Algae Competition in 2010 to envision new algae food and energy systems for our future. Presents at conferences, writes for industry media and has produced over 20 educational videos. Voted Algae Ambassador in Algae Industry Magazine 2015 Readers Poll.

Robert is CEO of Smart Microfarms, developing scalable microalgae systems for community, urban and rooftop farms and for algae entrepreneurs to grow high value algae food products that are local, sustainable and profitable. Marketing SpiruSource® products.

Publications: *Earth Food Spirulina* (1989) translated into 7 international editions. *Spirulina World Food* (2010, 2021). *Bamboo Architecture* (2011). *Imagine Our Algae Future* (2012) based on the International Algae Competition. *Algae Microfarms* (2013).

Robert Henrikson, PO Box 70124, Richmond CA 94807 USA.
roberthe@sonic.net.
spirulinasource.com • smartmicrofarms.com • algaecompetition.com
Facebook pages: AlgaeMicrofarms • AlgaeCompetition

Bibliography and references

1. Rediscovery of a 3.5 billion year old lifeform

1. Swimme, Brian. *The Universe is a Green Dragon.* Bear & Company, Santa Fe, NM, 1984, p. 137-8.
2. Kavaler, Lucy. *Green Magic: Algae Rediscovered.* Thomas Crowell, NY, 1983, p. 99-101.
3. Kavaler, p. 109-110.
4. Kavaler, p. 95-97.
5. Jassby, Alan. *Spirulina: a model for microalgae as human food.* Algae and Human Affairs. Cambridge University Press, 1988, p. 152.
6. Ciferri, Orio. *Spirulina, the Edible Organism.* Microbio. Reviews, Dec. 1983, p. 572.
7. Switzer, Larry. *Spirulina, The Whole Food Revolution.* Bantam, NY, 1982, p. 12.

2. A nutrient rich super food for super health

1. FDA Talk Paper, No. 41,160, June 23, 1981, US Food and Drug Admin.
2. Jassby, Alan. *Spirulina: a model for microalgae as human food.* Algae and Human Affairs. Cambridge University Press,1988, p. 158.
3. Shimamatsu, H. Personal communication, May 8, 1989.
4. Kataoka, N. and Misaki, A. *Glycolipids isolated from spirulina maxima.* Agric. Biol. Chem. 47 (10), 2349-2355, 1983.

4. New research reveals health benefits

Spirulina and Covid-19, HIV, and Immune System

1. Erik Reynolds and Leonard Lerer. *Natural Product Therapeutics and COVID-19: The Case for Clinical Development of Spirulina Extracts.* AlgaeSciences.com. May 2020. USA
2. Liji Thomas, MD. *Could microalgae-derived antiviral compounds combat SARS CoV-2 and other viruses?* 2021. India.

Potent anti-viral activity

1. Ayehunie, S., Belay, A. et al. *Inhibition of HIV-1 replication by an aqueous extract of spirulina platensis (Arthospira platensis).* Journal of Acquired Immune Deficiency Syndromes and Human Retrovirology, 18:7-12. USA 1998.
2. Hayashi, T. et al. *Calcium Spirulan, an inhibitor of enveloped virus replication, from a blue-green alga spirulina.* Am. Chemical Soc. and Am. Soc. of Pharmacognosy. Journal of Natural Products, 1996. Vol 59, No. 1, p. 83-87.
3. Hayashi, K. et al. *An extract from spirulina is a selective inhibitor of herpes simplex virus Type 1 penetration into HeLa cells.* Phytotherapy Research, Vol. 7, p.76-80, 1993.
4. Hayashi, O. et al. *Enhancement of antibody production in mice by dietary spirulina.* J. of Nutritional Sciences and Vitaminology, 40, p. 431-441, 1994.
5. Dora Allegra Carbone et al. *Evaluation of Microalgae Antiviral Activity and Their Bioactive Compounds.* Pub. in Antibiotics 2021, 10, 746.
6. Ronald Henson and Richard Kozlenko DPM, PhD. MPH. *Spirulina effects on AIDS virus, cancer and the immune system. 1997.*

Spirulina benefits for HIV patients

1. Frank Stephane Winter et al. *The Effect of Arthrospira platensis Capsules on CD4 T-Cells and Antioxidative Capacity in a Randomized Pilot Study of Adult Women Infected with Human Immunodeficiency Virus Not under HAART in Yaoundé, Cameroon.* Pub. in JNutrients 2014, 6, 2973-2986; Cameroon.

Spirulina extract may prevent serious Covid-19

1. Azaf Tzachor et al. *Spirulina Potential Implications for Covid-19 Related Cytokine Storm Therapy.* pub in Marine Biotechnology (2021) 23:149–155. Feb 2021. Israel.

Heart health and cholesterol reduction

1. Nayaka, N. et al. *Cholesterol lowering effect of spirulina.* Tokai Univ, Japan. Nutrition *reduction of serum cholesterol.* Tokai Univ. Progress in Med. Nov. 1986, Vol 6, No. 11.
2. Becker, E.W. et al. *Clinical and biochemical evaluations of spirulina with regard to its application in the treatment of obesity.* Inst. Chem. Pfanz. Nutrition Reports International, April 1986, Vol. 33, No. 4, p. 565.
3. Devi, M.A. and Venkataraman, L.V. *Hypocholesterolemic effect of blue-green algae spirulina platensis in albino rats.* Nutrition Reports International, 1983, 28:519-530.
4. Kato, T. and Takemoto, K. *Effects of spirulina on hypercholesterolemia and fatty liver in rats.* Saitama Med. College, Japan. Japan Nutr Foods Assoc. Jour. 1984, 37:321.
5. Iwata, K. et al. *Effects of spirulina on plasma lipoprotein lipase activity in rats.* Journal Nutr. Sci. Vitaminol. 1990, 36:165-171.

Anti-aging and neuroprotective effects

1. Vila, Jennifer et al. *Spirulina, Aging, and Neurobiology. in Spirulina in Human Nutrition and Health ed. by* Belay, Amha and Gershwin M.E. CRC Press, Boca Raton FL. 2008. p 277.
2. Gemma, C. et al. *Diets enriched in foods with high antioxidant activity reduce age-induced decreases in cerebellar beta-adrenergic function.* J. Neurosci. 2002, 22 (14), 6114-6120.
3. Audrey M. Neyrinck et al. *Spirulina Protects against Hepatic Inflammation in Aging: An Effect Related to the Modulation of the Gut Microbiota?* Pub. in Nutrients Jun 2017. Belgium.
4. Vila, Jennifer et al. *Spirulina, Aging, and Neurobiology* p 285.

Beta carotene for antioxidant protection, healthy eyes

1. Seshadri, C.V. *Large Scale Nutritional Supplementation with spirulina alga.* All India Project. Shri Amm Murugappa Chettiar Research Center (MCRC) Madras. 1993.
2. Peto, R. et al. *Can dietary beta carotene materially reduce human cancer rates?* Nature, 1981, 290:201-208.
3. Shekelle, R.B. et al. *Dietary Vitamin A and risk of cancer in the Western Electric study.* Lancet, 1981, 8257: 1185-1189.

Reducing risks of cancers

1. Menkes, et al. *Serum beta carotene, vitamins A and E, selenium, and the risk of lung cancer.* Johns Hopkins. N.E. Journal of Medicine, Nov. 1986, p. 1250.
2. Blot, William. *Journal of the National Cancer Institute.* Sept. 15, 1993.
3. National Research Council. *Diet, Nutrition and Cancer.* National Academy Press, Washington DC, 1982.
4. Ben Amotz, A. Presentation to *Polysaccharides from microalgae workshop,* Duke Univ, 1987.

Anti-cancer tumor effects

1. Schwartz, J., Scklar, G., Suda, D. *Inhibition of experimental oral carcinogenesis by topical beta carotene.* Harvard School of Dental Med. Carcinogenesis, May 1986, 7(5) 711-715.
2. Schwartz, J., Scklar, G., Suda, D. *Growth, inhibition and destruction of oral cancer cells by extracts from spirulina.* Cancer & Nutrition, 6/88.
3. Babu, M. et al. *Evaluation of chemoprevention of oral cancer with spirulina.* Nutrition and Cancer V. 24, No. 2, p.197-202, 1995.

Phycocyanin builds blood

1. Zhang Cheng-Wu, et. al. *Effects of polysaccharide and phycocyanin from spirulina on peripheral blood and hematopoietic system of bone marrow in mice.* Second Asia-Pacific Conf. Ibid, April, 1994.
2. Evets, P. et. al. *Means to normalize the levels of immunoglobulin E, using the food supplement spirulina.* Grodenski State Medical Univ. Russian Federation Commitee of Patents and Trade. Patent (19)RU (11)2005486. Jan. 15, 1994.

Phycocyanin enhances the immune system

1. Iijima, N., Shimamatsu, H., et al. (inventors; Dainippon Ink & Chemicals assignee). *Anti-tumor agent and method of treatment therewith.* US patent pending, ref. P1150-726-A82679, App. 15 Sep. 1982.
2. Dainippon Ink & Chemicals and Tokyo Kenkyukai (inventors and assignee). *Anti-tumoral agents containing phycobilin- also used to treat ulcers and hemorrhoidal bleeding.* 1983, JP 58065216 A 830418.

Polysaccharides enhance the immune system

1. Besednova, T. et. al. *Immunostimulating activity of lipopolysaccharides in blue-green algae.* Zhurnal Mikrobiologii, Epidemiologii, Immunobiologii, 56 (12), p. 75-79. 1979.
2. Baojiang, G. et. al. *Study on effect and mechanism of Polysaccharides of spirulina on body immune function improvement.* Second Asia-Pacific Conference on Algal Biotechnology. Singapore, April, 1994, p. 24.
3. Zhang Cheng-Wu, et. al. *Effects of polysaccharide and phycocyanin from spirulina on peripheral blood and hematopoietic system of bone marrow in mice.* Second Asia Pacific Conf. Ibid, April, 1994.
4. Lisheng, L. et. al. *Inhibitive effect and mechanism of polysaccharide of spirulina on transplanted tumor cells in mice.* Marine Sciences, Qindao, China. N.5, 1991, p.33-38.
5. Hayashi, T. et al. *Calcium Spirulan, an inhibitor of enveloped virus replication, from a blue-green alga spirulina.* Journal of Natural Products, 1996. Vol 59, No. 1, p. 83-87.
6. Qureshi, M. et al. *Spirulina extract enhances chicken macrophage functions after in vitro exposure.* J. Nutritional Immunology, V.3(4) 1995, p. 35-45.
7. Qureshi, M. et al. *Dietary spirulina enhances humoral and cell-mediated immune functions in chickens.* Immunopharmacology and Immunotoxicology, 1996 (submitted).
8. Qureshi, M. et al. *spirulina exposure enhances macrophage phagocytic function in cats.* Immunopharmacology and Immunotoxicology, 1996 (submitted).
9. Belay, A., Henson, R., Ota, Y. *Potential Pharmaceutical Substances from Aquaculturally Produced Spirulina.* Earthrise Farms, Calipatria CA. Presented to World Aquaculture Society, New Orleans. Jan. 1994.
10. Ayehunie, S., Belay, A. et al. *Inhibition of HIV-1 replication by an aqueous extract of spirulina platensis (Arthospira platensis).* Journal of Acquired Immune Deficiency Syndromes and Human Retrovirology, 18:7-12. USA 1998.

Sulfolipids stop the AIDS virus

1. Gustafson K, et al. *AIDS-Antiviral sufolipids from cyanobacteria (blue-green algae).* Journal of the National Cancer Institute, August 16, 1989, p 1254.
2. Kataoka, N. and Misaki, A. *Glycolipids isolated from spirulina maxima: structure and fatty acid composition.* Agric. Biol. Chem. 47 (10), 2349-2355, 1983.
3. Venkataraman, L.V. and Becker, E.W. *Biotechnology & Utilization of Algae- The Indian Experience.* Sharada Press, Mangalore, India, 1985, p 114-115.
4. Boyd, M, Gustafson, K. et al. *Cyanovirin-N, a novel HIV-inactivating protein that targets viral GP120.* Am. Soc. of Pharmacognosy, UC Santa Cruz. July 27-31, 1996.

Reduces kidney poisons from mercury and drugs

1. Yamane, Y. *The effect of spirulina on nephrotoxicity in rats.* Presented at Annual Sym. of the Pharmaceutical Society of Japan, Apr. 15, 1988, Chiba Univ, Japan.
2. Fukino, H. et al. *Effect of spirulina on the renal toxicity induced by inorganic mercury and cisplatin.* Eisei Kagaku. 1990, 36:5.
3. Morita, T. et al. *Effect of chlorella, spirulina and chlorophyllin on fecal excretion of polychlorinated dibenzo-p-dioxins in rats.* Fukuouka Institute of Health and Environmental Studies. Japan Journal Toxicology Environ. Health. 43 (1) 42-47 1997.

Builds healthy lactobacillus

1. Tokai, Y., et al. *Effects of spirulina on caecum content in rats.* Chiba Hygiene College Bulletin (Japan), Feb. 1987, Vol. 5, no. 2.
2. Archer, D.L. and Glinsmann, W.H. *Intestinal infection and malnutrition initiate AIDS.* US FDA. Nutrition Research, 1985. 5:19-19. Archer, D.L. and Glinsmann, W.H. *Enteric infections and other cofactors in AIDS.* Immunology Today, 1985, Vol. 6, no.10.

Wound healing and antibiotic effects

1. Martinez-Nadal, N.G. *Antimicrobial activity of spirulina.* Paper presented at X Intl Cong. of Microbiology, Mexico City, Aug. 1970.
2. Jorjani, G., Amirani, P. *Antibacterial activities of spirulina platensis.* Maj. limy Puz. Danisk.Jundi Shap, 1978, 1:14-18.
3. Clement, G. et al. (inventors; Institute Francais de Petrol, assignee.) *Wound treating medicaments containing algae.* Fr. M. 5279 (Int. Cl. A61k), 11 Sep. 1967.
4. Yoshida, R. (inventor). *Spirulina hydrolyzates for cosmetic packs.* Japan. Kokai 77 31, 838 (Int. Cl. A61k7100), 10 Mar. 1977.

Effects against diabetes and hypertension

1. Takai, Y. et al. *Effect of water soluble and insoluble fractions of spirulina over serum lipids and glucose resistance of rats.* J. Jap. Soc. Nutr. Food Sci. 1991, 44: 273-277.
2. Iwata, K. et al. *Effects of spirulina on plasma lipoprotein lipase activity in rats.* Journal Nutr. Sci. Vitaminol. 1990, 36:165-171.

Iron bioavailability and correction of anemia

1. Johnson P., Shubert, E. *Availability of iron to rats from spirulina, a blue-green alga.* Nutrition Research, 1986, Vol. 6, 85-94.
2. Takemoto, K. *Iron transfer from spirulina to blood in rats.* Saitama Med. Col., Japan, 1982.
3. Takeuchi, T. *Clinical experiences of administration of spirulina to patients with hypochronic anemia.* Tokyo Medical and Dental Univ., Japan, 1978.

Athletic endurance and immune deficit

1. Trojacanec, Z. et al. *Influence of extensive training on the number of erythrocytes and hemoglobin level and its correction.* Inst. Med. Physiology, Skopje, Macedonia. Pub in XXIV FIMS World Congress of Sports Medicine, June 1998.
2. Juszkiewicz A. et al. *An attempt to induce an immunomodulatory effect in rowers with spirulina extract.* Journal of the International Society of Sports Sutrition 2018, 15-9.

Benefits for malnourished children

1. Ramos Galvan, R. *Clinical experimentation with spirulina.* Colloque sur la valeur nutritionelle des algues spirulines, Rueill, May 1973. Nat. Inst. of Nutrition, Mexico City.
2. Sautier, C. and Tremolieres, J. *Food value of spirulina in humans.* Ann. Nutrition Alim, 1976, 30:517-534. (French).

3. Fox, Ripley D. *Village health and energy system, Farende, Togo.* April 4,1986 letter. Fox, R. D. Algoculture: Spirulina, hope for a hungry world. Pub. by Edisud, France, 1986.
4. Seshadri, C.V. *Large Scale Nutritional Supplementation with spirulina alga.* All India Project. Shri Amm Murugappa Chettiar Research Center (MCRC) Madras. 1993.
5. Annapurna, V. et al. *Bioavailability of spirulina carotenes in preschool children.* National Institute of Nutrition. Hyderabad, India. J. Clin. Biochem Nutrition. 10 p. 145-151. 1991.
6. Fica, V. et al. *Observations on the utilization of spirulina as an adjuvant nutritive factor in treating some diseases accompanied by a nutritional deficiency.* Clinica II Medicala, Spitalui Clinic Municipiului, Bucuresti. Med Interna 36 (3), 1984.
7. Miao Jian Ren. *Spirulina in Jiangxi China.* Academy of Agricultural Science, Jiangxi province, China. Paper presented at Soc. Appl. Algology, Lille, France, Sep. 1987.
8. Yonghuang, W. et al. *The study on curative effect of zinc containing spirulina for zinc deficient children.* 5th Int'l Phycological Congress, Qingdao, China. June 1994.

Reduces radiation effects for Children of Chernobyl

1. Belookaya, T. *Correspondence from the Chairman of Byelorussian Committee "Children of Chernobyl"* May 31, 1991.
2. Loseva, L.P. and Dardynskaya, I.V. *Spirulina- natural sorbent of radionucleides.* Research Institute of Radiation Medicine, Minsk, Belarus. Presented at the 6th Intl Congress of Applied Algology, Czech Republic, Sep. 9, 1993.
3. Sokolovskiy, V. *Corres. from the First Sec. BSSR to the United Nations,* May 20, 1991.

Russian patent for radiation sickness

1. Evets, P. et. al. *Means to normalize the levels of immunoglobulin E, using the food supplement spirulina.* Grodenski State Medical Univ. Russian Federation Commitee of Patents and Trade. Patent (19)RU (11)2005486. Jan. 15, 1994.
2. Loseva, L. P. *Spirulina platensis and specialties to support detoxifying pollutants and to strengthen the immune system.* Research Inst. of Radiation Medicine, Minsk, Belarus. presented at 8th Int'l Congress Applied Algology, Italy. Sep. 1999.

GLA and prostaglandin stimulation

1. Tudge, C. *Why we all need the evening primrose.* New Scientist, Nov. 1981, 506:23.
2. Kunkel, S.L. et al. *Suppression of chronic inflammation by evening primrose oil.* Progress in Lipids, 1982, Vol. 20, p. 885-888.
3. Kernoff, P.B.A, et al. *Antithrombotic potential of DGLA.* Brit Med. J., 1977, 2:1441-1444.
4. Vadaddi, K.S., Horrobin, D.F. *Weight loss produced by evening primrose oil administration.* IRSC Med. Sci., 1979, 7:52.
5. Huang, Y.S. et al. *Most biological effects of zinc deficiency corrected by GLA.* Atheroscelosis, 1982, 41:193-208.
6. Horrobin, D.F. *The possible roles of prostaglandin E1 and of essential fatty acids in mania, depression and alcoholism.* Prog.in Lipids, 1981. Vol 20, 539-541. Horrobin, *Loss of delta-6-desaturase activity in aging.* Med. Hypotheses, 1981, 7:1211-1220.
7. Nichols, B., Wood, B. *The occurrence and biosynthesis of gamma linolenic acid in spirulina platensis.* Lipids, 1986, Vol 3, No. 1, 46-50.

Weight loss research

1. Becker, E.W. et al. *Clinical and biochemical evaluations of the alga spirulina with regard to its application in the treatment of obesity.* Inst. Chem. Pflanzenphysiologie. Nutrition Reports International, April 1986, Vol. 33, No 4, 565.

Algae medicines from bioactive compounds

1. Edwards, Mark R. *Emerald Renaissance.* 2020. Amazon.com.

5. The variety of products around the world

1. *A safe diet pill, You'll never go hungry.* National Enquirer, June 1, 1981.
2. *Linablue-A (Natural blue colorant of spirulina origin) Technical Information.* Dainippon Ink & Chemicals, Tokyo, Japan, 1985.
3. Henson, R. *Spirulina algae improves Japanese fish feeds.* Aquaculture Magazine. Nov/Dec 1990. V. 6 N.6. p 38-43.
4. Henson, R. *Health food for the Aquarium.* Fresh and Marine Aquarium.1993. p 70-72.
5. Henson, R. *Super-Foods for birds.* American Cage-Bird Mag. April, 1993. p. 48-51.
6. Qureshi, M. et al. *Immune enhancement potential of spirulina platensis in chickens.* Dept. of Poultry Science, NC State Univ. Raleigh NC. Poultry Sci. Assoc, Aug. 1994.
7. Nawrocka D, et al. *Spirulina improves mitochondrial function impaired by elevated oxidative stress... in Equine Metabolic Syndrome horses.* Mar Drugs 2017 Aug 3:15(8).
8. Kronik, M. andcka D. et al. *Immunoassay techniques with fluorescent phycobiliprotein conjugates.* Clin Chem., 1983, 29:9,1582-86.
9. *A better way to track diseases.* Chemical Week, August 31, 1985.
10. Kawamura, M. et al. 1986. *A new restriction enzyme from spirulina platensis.* Nucleic Acids Research, 1986, Vol. 14, No. 5.
11. Shimamatsu, Hidenori. Personal communication, 1988.

6. How spirulina is ecologically grown

1. Earthrise Nutritionals California USA. earthrise.com.
2. June Pharmaceuticals Myanmar. junespirulina.com.
3. Cyanotech Hawaii USA. cyanotech.com.
4. Boonsom Farm Thailand. boonsomfarm.com.
5. Parry Nutraceuticals India. parrynutraceiticals.com.
6. Bluetec Naturals Inner Mongolia China. bestphycocyanin.com.
7. Australian Spirulina Australia. australianspirulina.com.au.
8. Jongerius Ecoduna Austria. jongerius-ecoduna.at.
9. Agcore Technologies Rhode Island USA. agcoretech.com.
10. Spiruline La Capitelle France. spirulinelacapitelle.com.
11. Federation des Spiruliniers de France. spiruliniersdefrance.fr.
12. La Spiruline à l'Eco-domaine de Bouquetot France. hyes.eu.
13. La Spirulina de Haute Saintonge France. spiruline-fr.com.
14. Energaia Thailand. energaia.com.
15. Spirulina Viva Mexico. espirulinaviva.org.
16. Smart Microfarms California USA. smartmicrofarms.com.

7. Spirulina in the developing world

1. U.N. World Health Organization, Geneva, Switzerland. Correspondance. June 8, 1993.
2. Switzer, Larry. *Spirulina- The Whole Food Revolution.* Bantam, NY, 1982, p.115-116.
3. Association Pour Combattre la Malnutrition par Algoculture (ACMA). A non-profit organization. 34190 St. Bauzille-de-Putois, France.
4. Fox, Ripley D. *Algoculture: Spirulina, hope for a hungry world.* Edisud, Aix-en Provence, France, 1986.
5. Fox, Ripley D. 1984. *Spirulina- the alga that can end malnutrition.* Futurist, Feb. 1985.
6. Seshadri, C.V. *Large Scale Nutritional Supplementation with spirulina alga.* All India Project. Shri Amm Murugappa Chettiar Research Center (MCRC) Madras. 1993.
7. Seshadri, C.V. and Jeeji Bai, N. *Spirulina Nat. Sym.*. MCRC, Madras, India 1992.
8. Mahamat Sorto. sortoma@yahoo.fr.

9. Antenna Foundation Geneva Switzerland. antenna.ch.
10. Antenna France. antenna-france.org.
11. Spirulina Nutritech Foundation India. spirufoundation.org.
12. Aurospirul Farms Auroville India. aurospirul.com.
13. June Pharmaceuticals Myanmar. junespirulina.com.
14. Nguyen Lan Dinh. 1990. In Fox, Denise. *Health Benefits of Spirulina.* p 183.
15. Bucaille, P. 1990. In Fox, Denise . *Health Benefits of Spirulina.* p 183.
16. Picard, M.E. 1993. In Fox, Denise. *Health Benefits of Spirulina.* p 182.
17. Brouillat, E. 1993. In Fox, Denise. *Health Benefits of Spirulina.* p 182.

8. Algae's impact on food, agriculture and climate

1. *An assessment of the resource base that supports the global economy.* World Resources Institute, Basic Books, New York, 1988, p. 4.
2. Higgins, et. al. *Potential population supporting capacities of lands in the developing world.* Technical Report of Project FPA/INT/513. FAO, United Nations, Rome, 1983.
3. Postel, Sandra. *Land's End.* Worldwatch , May-June 1989, p. 13.
4. Repetto, Robert, et. al. *Wasting Assets. Natural Resources in the National Income Accounts.* World Resources Institute. 1989.
5. Ota, Yochimichi. Earthrise Farms. Personal communication, 1988.
6. Allen, Mark. BiofuelsDigest.com/bdigest/2020/10/29/terrasync-algae-micro-cover-crop/
7. news.algaeworld.org/2019/04/all-hail-spirulina-the-most-super-of-superfoods/
8. 3dprintingindustry.com/news/researchers-achieve-cereal-manufacturing-via-3d-printed -algae-based-healthy-snacks-178204
9. asianscientist.com/2021/04/tech/microalgae-burger-singapore-sophies-bionutrients/10. food ingredientsfirst.com/news/heme-20-spirulina-extract-challenges-impossible-foods -bleeding-hero-ingredient-in-plant-patties.html
11. *Blue-green algae for rice production.* FAO of the U.N FAO Soils Bulletin. Rome, 1981.
12. Edwards, Mark R. *Emerald Renaissance 2019 p 182 .* amazon.com. *Algae Biosciences Bio Max,* 2011.
13. Edwards, Mark R. *Emerald Renaissance 2019 p 118.* amazon.com.
14. Tugce Senturk Parreidt et al. *Alginate-Based Edible Films and Coatings for Food Packaging Applications. Foods 2018, 7, 170.*
15. Puganeswary Sukumaran et al. *Potential of fresh POME as a growth medium in mass production of Arthrospira platensis. Int.J.Curr.Microbiol.App.Sci (2014) 3(4): 235-250.*
16. Tuomas Huokko et al. *Probing the biogenesis pathway and dynamics of thykaloid membranes. Nature Communications. doi.org/10.1038/s41467-021-23680-1.*

9. Microalgae's role in restoring our planet

1 Mathewson S. *Algae 'Bioreactor' on Space Station Could Make Oxygen, Food for Astronauts.* space.com/space-station-algae-experiment-fresh-air.html. May 2019.
2. *Astronaut Alexander Gerst checks a culture bag for the Space Algae experiment.* Aug 20, 2018. Nasa.gov.
3. Fahrion, J. et al. *Use of Photobioreactors in Regenerative Life Support Systems for Human Space Exploration.* Frontiers in Microbiology. 29 Jun 2021 doi: 10.3389 fmicb.2021.699525
4. Illustration from O'Callahan, J. *How to make oxygen on Mars.* Mail online 15 May 2015.
5. Henrikson R., Edwards M. *Imagine Our Algae Future.* p 79. 2012. Amazon.com.
6. Henrikson R., Edwards M. *Imagine Our Algae Future.* p 128. 2012. Amazon.com.
7. Fox, Ripley. *Spirulina, production & potential.* Edisud. Aix-en-Provence, France. 1996.
8. news.algaeworld.org/2014/08/floating-bed-method-proven-feasible-spirulina-farming method-malaysia-2/

9. Federation des Spiruliniers de France. spiruliniersdefrance.fr.
10. Fox, Ripley. *Spirulina, production & potential.* Edisud. Aix-en-Provence, France. 1996.
11. Jourdan, Jean-Paul. *Cultivez Votre Spiruline - Manuel de Culture Artisanale de la Spiruline.* 1999. Jean-Paul Jordan, Le Castanet, Mialet, 30140 Anduz, France. Pub. by Antenna Technologie, 1999.
12. Nancy Jack Todd and John Todd. *Bioshelters, Ocean Arks and City Farming.* Sierra Club Books, San Francisco, 1984, p. 118.
13. Edwards, Mark. *Peace Microfarms A Green Strategy to Prevent War.* 2017. Amazon.com.
14. *Breakthrough biological innovation - the key to profitable and sustainable vertical farming* Back of the Yards Algae Sciences. algaesciences.com. Nov 2020.
15. Buehrer Mark. *Net Zero Urban Farm* - 2020engineering.com. 2020.
16. *Algae greenhouse neutralizes carbon footprint data centre* - hortidaily.com/article/9308296/algae-greenhouse-neutralizes-carbon-footprint-data-centre/. Apr 2021.
17. Clive Thompson. *No Parking Here.* motherjones.com/environment/2016/01/future-parking-self-driving-cars/. Jan 2016.
18. Edwards, Mark R. *Ecolanda. the Frst Emerald Economy.* pg 18. 2021. Amazon.com.
19. Edwards, Mark R. *Emerald Renaissance.* 2020. Amazon.com.
20. Lovelock, James. *The Ages of Gaia.* W.W. Norton & Co. Inc., N.Y. 1988, p. 179.

Appendix A: Quality and safety standards

1. Delpeuch, F. et. al. *Consumption as food and nutritional composition of blue-green algae among populations in the Kanem region of Chad.* Ann. Nutr. Aliment. 29, 497-516. 1976.
2. Institut Francais du Petrol. *Rapport ou Comite Consultatif des Proteines* OAA/ OMS/ FISE-Etat d'Avancement du Procede IFP de Production d'algues Dec 1970, p. 11.
3. Fadoul, L. *Les algues bleues du Kanem.* Rapport de mission par L. Fadoul, A. Avrem et G. Le Guedes (experts de la division de la nutrition, FAO) Juin 1971.
4. Takemoto, K. *Subacute toxicity study with rats.* Saitama Medical College, Japan, 1982.
5. Atatsuka, K. *Acute toxicity and general pharmacological studies.* Meiji College of Pharmacy, Japan,1979.
6. Chamorro-Cevallos, G. *Toxicological research on the alga spirulina.* UNIDO, 24 Oct. 1980, UF/MEX/78/048. (French)
7. Becker, W. E., Vanattaraman, L.V. et al. *Production and utilization of the blue-green algae Spirulina in India.* Biomass, 4, 105-125, 1984.
8. Chamorro, G. et al. *Subchronic toxicity study in rats fed Spirulina.* J. de Pharmacie de Belgique 43, 29-36, 1988.
9. Salazar, M., Chamorro, G. et al. *Effect of spirulina consumption on reproduction and peri-and postnal development in rats.* Food and Chemical Toxicity 34, 353-359. 1996.
10. Waslein, C. et al. *Uric acid levels in men fed algae and yeast as protein sources.* J. Food. Sci. 1970, 35, 294-8.
11. Jassby, Alan. *Spirulina: a model for microalgae as human food.* Algae and Human Affairs, Cambridge Univ Press, 1988, p. 159.
12. Boudene, C. et al. *Evaluation of long term toxicity on rats with spirulina.* Ann Nutr. Aliment., 1976, 30: 577-588.
13. Til, H.P, and Williams, M. *Sub-chronic toxicity study with dried algae in rats.* Cent. Inst. for Nutrition and Food Research, Zeist, Ned., 1971.
14. Fevrier, C. and Seve, B. *Incorporation of spirulina into pig diets.* Ann. Nur. Aliment., 1976, 29: 625-30 (French).
15. Earthrise Farms. *Five year testing of heavy metals in spirulina, 1983-1987.* 1988.
16. Carmichael, W.W. *The toxins of Cyanobacteria.* Sci. American, Jan. 1994, p. 78-86.
17. An, J., and Carmichael, W.W. *Technical Booklet for the Microalgae Biomass Industry: Detection of microcystins and nodularins using an enzyme linked immunosorbant assay (ELISA).* 1996.

Spirulina and algae bookshelf

Algae Microfarms
Microfarms for community and urban gardens, rooftop, vertical farms and living buildings.
Robert Henrikson. 2013. amazon.com.

Imagine Our Algae Future
Visionary algae architecture and landscape designs.
Robert Henrikson, Mark Edwards. 2012. amazon.com.

Spirulina World Food
How this micro algae can transform your health and our planet.
Robert Henrikson. 2010. amazon.com.

Emerald Renaissance
Restore health for people, producers and our planet.
Mark R. Edwards. 2019. amazon.com.

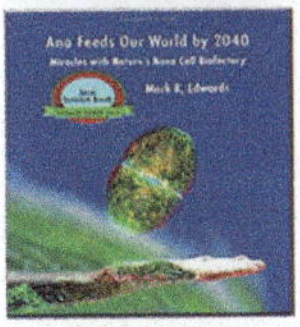

Ana Feeds Our World by 2040
Miracles with Nature's Nano Cell Biofactory.
Mark R. Edwards. 2017. amazon.com.

Peace Microfarms
A Green Algae Strategy to prevent War.
Mark R. Edwards. 2012. amazon.com.

Tiny Mighty Al
The tiny plant that saved our planet.
Mark R. Edwards. 2010. amazon.com.

Made in United States
North Haven, CT
04 November 2026

21760277R00124